D0496832

Tom Wolfe is the author of more than a dozen books, among them *The Electric Kool-Aid Acid Test*, *The Right Stuff*, *The Bonfire of the Vanities*, *A Man in Full*, *I Am Charlotte Simmons*, and *Back to Blood*. He received the National Book Foundation's 2010 Medal for Distinguished Contribution to American Letters. He lives in New York City.

TOM WOLFE

The Kingdom of Speech

VINTAGE

1 3 5 7 9 10 8 6 4 2

Vintage
20 Vauxhall Bridge Road,
London SW1V 2SA

Vintage is part of the Penguin Random House group of companies
whose addresses can be found at global.penguinrandomhouse.com

Penguin
Random House
UK

First published in Vintage in 2017
First published in hardback by Jonathan Cape in 2016

penguin.co.uk/vintage

A CIP catalogue record for this book is
available from the British Library

ISBN 9781784704896

Printed and bound by Clays Ltd, St Ives Plc

Penguin Random House is committed to a sustainable future
for our business, our readers and our planet. This book is made
from Forest Stewardship Council® certified paper.

MIX
Paper from
responsible sources
FSC® C018179

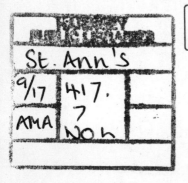

With a deep bow,
the author thanks
CHRISTINA VERIGAN
for such a big slice
of her erudition

THE KINGDOM OF SPEECH

CHAPTER I

THE BEAST WHO TALKED

ONE BRIGHT NIGHT in the year 2016, my face aglow with god-knows how many MilliGAUSS of x-radiation from the computer screen in front of me, I was surfing the net when I moused upon a web node reading:

THE MYSTERY OF LANGUAGE EVOLUTION*

It seems that eight heavyweight Evolutionists**—linguists, biologists, anthropologists, and computer scientists—had

---

* It was a scholarly article in *Frontiers in Psychology* ("The Mystery of Language Evolution," May 7, 2014, available at: dx.doi.org/10.3389/fpsyg.2014.00401).

** They were: Marc D. Hauser, Charles Yang, Robert C. Berwick, Ian Tattersall, Michael J. Ryan, Jeffrey Watumull, Noam Chomsky, and Richard C. Lewontin.

3

published an article announcing they were giving up, throwing in the towel, folding, crapping out when it came to the question of where speech—language—comes from and how it works.

"The most fundamental questions about the origins and evolution of our linguistic capacity remain as mysterious as ever," they concluded. Not only that, they sounded ready to abandon all hope of *ever* finding the answer. Oh, we'll keep trying, they said gamely...but we'll have to start from zero again. One of the eight was the biggest name in the history of linguistics, Noam Chomsky. "In the last 40 years," he and the other seven were saying, "there has been an explosion of research on this problem," and all it had produced was a colossal waste of time by some of the greatest minds in academia.

Now, that was odd...I had never heard of a group of experts coming together to announce what abject failures they were...

Very odd, in fact...so I surfed and Safaried and finally moused upon the only academic I could find who disagreed with the eight failures, a chemist at Rice University...*Rice*... Rice used to have a big-time football team...the Rice Owls...wonder how they're doing now? I moused around on the Rice site some more, and *uh-oh*...not so great last season, the Owls...*football*...and I surfed to *football concussions*...exactly as I thought! There's a regular epidemic of concussions raging! They're busy beating each other into

4

clots of early Alzheimer's!...*concussions*...surfing surfing surfing, but look at this! Football is nothing compared to *ice hockey*...without at least two concussions under your skull you aren't even ready for the NHL—

—and all the while something else was so caught on my pyramids of Betz that not even an NHL enforcer's head check could have dislodged it: they can't figure out what language *is*. One hundred and fifty years since the Theory of Evolution was announced, and they had learned...*nothing*...in that same century and a half, Einstein discovered the speed of light and the relativity of speed, time, and distance...Pasteur discovered that microorganisms, notably bacteria, cause an ungodly number of diseases, from head colds to anthrax and oxygen-tubed, collapsed-lung, final-stage pneumonia...Watson and Crick discovered DNA, the so-called building blocks genes are made of...and 150 years' worth of linguists, biologists, anthropologists, and people from every other discipline discovered...*nothing*... about language.

What is the problem? Speech is not one of man's several unique attributes—speech is the attribute of all attributes! Speech is 95 percent plus of what lifts man above animal! Physically, man is a sad case. His teeth, including his incisors, which he calls eyeteeth, are baby-size and can barely penetrate the skin of a too-green apple. His claws can't do anything but scratch him where he itches. His stringy-ligament body makes him a weakling compared to all the animals his size. Animals his size? In hand-to-paw, hand-to-claw, or hand-to-incisor combat, any animal his size would have him for lunch. Yet man owns or controls them all, every animal that exists, thanks to his super-power: speech.

What is the story? What is it that has left endless gener-ations of academics, certified geniuses, utterly baffled when it comes to speech? For half that time, as we will see, they formally and officially pronounced the question unsolvable and stopped trying. What is it they still don't get after a veritable eternity?

Our story begins inside the aching, splitting head of Alfred Wallace, a thirty-five-year-old, tall, lanky, long-bearded, barely grade-school-educated, self-taught British naturalist who was off—alone—studying the flora and the fauna of a volcanic island off the Malay Archipelago near the equa-tor...when he came down with the dreaded Genghis ague (rhymes with "bay view"), today known as malaria. So here he is, in not much more than a thatched hut, stretched out, stricken, bedridden, helpless...and another round of the paroxysms strikes with full force...the chills, the rib-rat-tling shakes...the head-splitting spike of fever followed by a sweat so profuse it turns the bed into a sodden tropical bog. This being 1858 on a miserable, sparsely populated speck of earth somewhere far, far south of London's nobs, fops, top hats, and toffs, he has nothing with which to while the time away except for a copy of *Tristram Shandy* he has already read five times—that and his own thoughts...

One day he's lying back on his reeking bog of a bed...thinking...about this and that...when a book he read a good twelve years earlier comes bubbling up his brain stem: *An Essay on the Principle of Population* by a Church of England priest, Thomas Malthus.*

The priest had a deformed palate that left him with a speech defect, but he could write like a dream. The book had been published in 1798 and was still very much alive sixty years and six editions later. Left unchecked, Malthus said, human populations would increase geometrically, doubling every twenty-five years.[1] But the food supply increases only arithmetically, one step at a time.[2] By the twenty-first century, the entire earth would be covered by one great heaving mass of very hungry people pressed together shank to flank, butt to gut. But, as Malthus predicted, something *did* check it—namely, Death, unnatural Death in job lots...starting with starvation, vast famines of it...disease, vast epidemics of it...violence, mayhem, organized slaughter, wars and suicides and gory genocides...to the cantering hoofbeats of the Four Horsemen culling the herds of humanity until but a few, the strongest and the healthiest, are

---

* This experience is recounted by Ernest H. Rann, who interviewed Wallace for the article "Dr. Alfred Russel Wallace at Home," *The Pall Mall Magazine* (March 1909).

left with enough food to survive. This was precisely what happened with animals, said Malthus.

Ahura! It lights up Wallace's brainpan with a flash—*It!*—the solution to what naturalists called "the mystery of mysteries": how Evolution works! Of course! Now he can see it! Animal populations go through the same die-offs as man. All of them, from apes to insects, struggle to survive, and only the "fittest"—Wallace's term—make it. Now he can see an inevitable progression. As generations, ages, eons go by, a breed has to adapt to so many changing conditions, obstacles, and threats that it turns into something else entirely—a *new* breed, a new *species!*—in order to survive.

For at least sixty-four years British and French naturalists, starting with the Scotsman James Hutton[3] and the Englishman Erasmus Darwin[4] in 1794 and the Frenchman Jean-Baptiste Lamarck in 1800,[5] had been convinced that all the various species of plants and animals of today had somehow evolved from earlier ones. In 1844 the idea had lit up the sky in the form of an easy-to-read bestselling book called *Vestiges of the Natural History of Creation,* a complete cosmology of the creation of earth and the solar system and plant and animal life from the lowliest forms up through the transmutations of monkeys into man. It transfixed readers high and low—Alfred, Lord Tennyson; Gladstone, Disraeli, Schopenhauer, Abraham Lincoln, John Stuart Mill; and Queen Victoria and Prince Albert (who read it aloud to each other)...as well as the general public—in droves.

There was no author's name on the cover or anywhere in its four hundred pages. He or she—there were those who assumed a writer this insidious must be a woman; Lord Byron's too-clever-by-half daughter Ada Lovelace was one suspect—apparently knew what was coming.[6] The book and Miss, Mrs., or Mr. Anonymous caught Holy Hell from the Church and its divines and devotees. One of the pillars of the Faith was the doctrine that Man had descended from Heaven, definitely not from monkeys in trees. Among the divines, the most ferocious attack was the Reverend Adam Sedgwick's in the *Edinburgh Review*.[7] Sedgwick was an Anglican priest and prominent geologist at Cambridge. If words were flames, Sedgwick's would have burned the anonymous heretic at the stake. The miserable creature gave off the stench of "inner deformity and foulness."[8] His mind was hopelessly twisted with "gross and filthy views on physiology,"[9] if indeed he still had a mind. The foul wretch thought that "religion is a lie" and "human law a mass of folly, and a base injustice," and "morality is moonshine." In short, this disgusting apostate thought "he can make man and woman far better by the help of a baboon" than by the mercies of the Lord God.[10]

Then the book caught high-IQ hell from the ranks of established naturalists in general. They found it journalistic and amateurish; which is to say, the work of an unknown outsider and a threat to their status. The boy wonder of the "serious" scientific establishment, Thomas Henry Huxley,

9

then twenty-eight, wrote what was later described as "one of the most venomous book reviews of all time"[11] when the tenth edition came out in 1853. He called *Vestiges* "a once attractive and still notorious work of fiction."[12] As for its anonymous author, he was one of those ignorant and superficial people who "indulge in science at second-hand and dispense totally with logic."[13] Everyone in the establishment was happy to point out that this anonymous know-it-all couldn't begin to explain how, through what physical process, all this transmutation, this evolution, was supposed to have taken place. Nobody could figure it out—until *now,* a few *moments* ago, inside *my* brain! *Mine!* Alfred Russel Wallace's!

He is still in his wet, reeking bed, trying to endure the endless malarial paroxysms, when another kind of fever, an *exhilarating* fever, seizes him...a fervid desire to record his revelation and show the world—*now!* For two days and two nights[14]...during every halfway tranquil moment between the chills, the rattling ribs, the fevers, and the sweats...he writes and he writes writes writes a twenty-plus-page manuscript entitled "On the Tendency of Varieties to Depart Indefinitely from the Original Type."[15] He has done it! His will be the first description ever published of the evolution of the species through natural selection. He sent it off to England on the next boat...

...but not to any of the popular scholarly publications, such as the *Annals and Magazine of Natural History* and *The*

Literary Gazette; and Journal of Arts, Belles Lettres, Sciences, &c., where he had published forty-three papers during his eight years of field work in the Amazon and here in Malay. No, for this one—for *It*—he was going to mount the Big Stage. He wanted this one to go straight to the dean of all British naturalists, Sir Charles Lyell, the great geologist. If Lyell found merit in his stunning theory, he had the power to introduce it to the world in a heroic way.

The problem was, Wallace didn't know Lyell. And on this primitive little island, where was he going to get his address? But he had corresponded a few times with another gentleman who was a friend of Lyell's, namely, Erasmus Darwin's grandson Charles. Charles had happened to mention in a letter two years earlier, in 1856, that Lyell had praised one of Wallace's recent articles (probably "On the Law Which Has Regulated the Introduction of New Species," also known as the "Sarawak Law" paper, 1855).[16] By early March of 1858, Wallace's manuscript and a letter were on the ocean, 7,200 miles from England, addressed to Charles Darwin, Esq. Exceedingly polite was that letter. It all but cringed. Wallace was asking Darwin to please read his paper and, if he thought it worthy, to please pass it along to Lyell.

So it was that Wallace put his discovery of all discoveries—the origin of species by natural selection— into the hands of a group of distinguished British Gentlemen. The year 1858 was on the crest of the high Victorian tide of the British Empire's dominion over palm and pine. Britain was the most powerful military and economic power on earth. The mighty Royal Navy had seized and then secured colonies on every continent except for the frozen, human-proof South Pole. Britain had given birth to the Industrial Revolution and continued to dominate it now, in 1858, almost a century later. She controlled 20 percent of all international trade and 40 percent of all industrial trade. She led the world in scientific progress, from mechanical inventions to advances in medicine, mathematics, and theoretical science.

To give a face to all that, she had at her disposal the most highly polished aristocrat in the West...the British Gentleman. He might or might not have a noble title. He might be a Sir Charles Lyell or a Mr. Charles Darwin. It didn't matter. Other European aristocrats, even some French ones, lifted their forearms before their faces to shield their eyes in the British Gentleman's presence. The gleam and refinement of the usual frills—manners, dress, demeanor, tortured accent, wit, and wit's lacerating weapon, irony—were the least of it. The most of it was wealth, preferably inherited.

The British Gentleman, better known in ages past as a member of the landed gentry, typically lived on inherited wealth upon a country estate of one thousand acres or more, which he rented out for farming by the lower orders.[17] He went to Oxford (Lyell) or Cambridge (Darwin) and might become a military officer, a clergyman, a lawyer, a doctor, a prime minister, a poet, a painter, or a naturalist—but he didn't *have* to do anything. He didn't have to work a day in his life. Sir Charles Lyell's ascension to the status of British Gentleman had begun the day his grandfather, also Charles Lyell, converted a naval career into enough money to buy an estate with endless acres and a palatial manor house in Scotland and retire as a high-living lairdly sort who was no longer hobbled socially by the need to work. His erstwhile naval career, which had been a necessity at the time, cast a bit of a shadow upon him, but his son (another) Charles was born free of that curse, and in due course his grandson became Sir Charles Lyell (third Charles in a row),[18] thanks to his achievements in geology.* The Darwin line went back much further than that, some two hundred years, to the mid-1600s, to Oliver Cromwell and his serjeant-at-law—i.e., lawyer—one Erasmus Earle.[19] Erasmus parlayed his position into a small fortune and huge landholdings, and never again did a gentleman in the ensuing eight or so Earle-Darwin generations have to work.

Charles Darwin's father, Robert Darwin, was a doctor—like *his* father, Erasmus Darwin. His true passion however, was investing, lending, brokering, betting, and otherwise dealing in the Industrial Revolution's money markets. He made an absolute fortune...then multiplied it by marrying

---

* Sir Charles Lyell was knighted in 1848 and made a baronet in 1864. He received the Copley Medal for his scientific work in 1858. He was buried in Westminster Abbey.

a daughter of Josiah Wedgwood, one of the early giants of industry. Wedgwood was a potter, a craftsman, who had created factories that produced chinaware finer than any plain potter ever dreamed of. Robert Darwin's arena was London and its financial district, the City. But like most great British Industrial Revolutionaries, he chose to live in the countryside on a large and largely irrelevant estate—his, in Shropshire, was called the Mount—to show that he was as grand as the landed gentry of yore.[20] He paid for his son Charles to study medicine at the University of Edinburgh (the boy dropped out), then sent him to Christ's College at Cambridge to become a clergyman (the boy dropped out), then had to settle for the boy dropping down to the bottom at Cambridge and barely getting a bachelor of arts degree (without honors or the vaguest idea of what to do with his life), and then begrudgingly paid for the boy to enjoy a five-year voyage of exploration, or sightseeing, or something, aboard a boat named for a dog, His Majesty's Ship *Beagle,* to prepare him for a career in the field of— as far as Dr. Darwin could tell—nothing. Once the boy was finished with that nonsense, Dr. Darwin, who himself had married a Wedgwood relative, nudged the boy, who was twenty-nine, into marrying in 1839 his perfectly nice, if plain, thirty-year-old spinster first cousin, Emma Wedgwood, and in 1842 bought them a country place, Down House, southeast of London, and settled enough money on him for the boy to live well forever and ever. Living well included eight or nine servants—a butler, a cook, a manservant or two, a parlor maid, a lady's maid, and at least one nanny and a governess—from day one.*

Where did this, the eternally Daddy-paid-for life of a British Gentleman, leave someone like Alfred Russel Wallace? His father, a lawyer, had undertaken a legal career and a business career—and a family and a half, namely, a wife and nine children (Alfred was the eighth)—and wound up swindled and bankrupt, utterly wiped out. The Wallaces were the very picture of what is known today as a downwardly mobile family. They didn't have the money for Alfred's education beyond grade school. Years later, to pay for his explorations in the Amazon and Malay, Wallace had to ship stupendous loads of (dead) snakes, mammals, shells, birds, beetles, butterflies—lots of flamboyant butterflies—moths, gnats, and no-see-um bugs to an agent in England...who sold them to scientists, amateur naturalists, collectors, butterfly lovers, and anyone else intrigued by exotic oddments from the underbellies of the earth. A single shipment might contain thousands of items. The kind of mortal willing or forced by Fate to go out into ankle-sucking muck, brain-frying heat, clouds of mosquitoes, ague-ghastly nightscapes...on terrains slithering with poisonous

---

* Emma Darwin wrote in her diary about hiring servants for her move to Down House. She also documented many details about Charles's health and their family life. Her complete diaries are available digitally at *Darwin Online* (Darwin-online.org.uk).

snakes...to harvest hundreds of curiosities at a time[21]...
was called a flycatcher. Gentlemen like Lyell and Darwin
didn't think of flycatchers as fellow naturalists but as sup-
pliers on the order of farmers or cottage weavers.

There you had Alfred Wallace...a flycatcher. The very
thought of *having to make a living* at all, much less as a
Malaysian bugmonger, was enough to set any gentleman to
itching and scratching all over...and in the mid-nineteenth
century, the Gentlemen ran every major area of British life:
politics, religion, the military, the arts and sciences. Wallace
was well aware that he was about to get in touch with a social stratum far above his own. But he wasn't writing Lyell,
via Darwin, seeking social acceptance. All he sought was
professional recognition by some eminent fellow naturalists.

How very naive of him! The British Gentleman was not
merely rich, powerful, and refined. He was also a slick oper-
ator...smooth...smooth...smooth and then some. It was
said that a British Gentleman could steal your underwear,
your smalls and skivvies and knickers, and leave you staring
straight at him asking if he didn't think it had turned rather
chilly all of a sudden.

When he received the manuscript and the letter, in June of
1858—and please forgive an anachronism, namely, a verb
from almost exactly one century later—Mr. Charles Darwin freaked out. He delivered the manuscript to his good

friend Lyell, all right...along with a bleating yelp for help. In twenty pages this man Wallace had forestalled his life's work—*his entire life's work!* "Forestalled" was the 1858 word for "scooped."

Darwin had achieved a solid reputation among naturalists with a series of monographs about coral reefs, volcanic islands, fossils, barnacles, the habits of mammals...he had written an engaging and highly praised book, *Journal of Researches,* about his five-year (1831–36) voyage aboard His Majesty's Ship *Beagle,* one of England's many government-sponsored worldwide explorations in the nineteenth century. He had been elected not only to the Geological Society but also to the most prestigious scientific body in England, the Royal Society of London, whose membership was restricted to eight hundred, namely, the eight hundred leading scientists in the world. Fine: and all that meant nothing to him in light of his Theory of Evolution, his very much *secret* life's work.

He had started thinking about evolution—"transmutation" was the term for it at the time—when he was on the *Beagle.* By 1837, a year after the expedition had ended, he was convinced that all plant and animal life on earth was the result of the transmutation, i.e., evolution, of all the various species over millions of years. And not just plants and animals...the *Beagle* explorers spent long intervals on land, and Darwin kept coming upon natives so primitive they struck a British Gentleman like himself as closer to

apes than to humans…particularly the Fuegians. The Fuegians (pronounced "*Fway*gians") were natives of Tierra del
Fuego, an Argentine and Chilean province so far south that
the tip at the bottom is part of Antarctica. The Fuegians
were brown and sun-wrinkled and hairy. The hair on their
heads was as wild as a howling…a howling…well, as a
howling hairy ape's. Their hairy legs were too short and
their hairy arms too long for their hairy torsos. In Darwin's eyes, the only thing that distinguished the Fuegians
from the higher apes was the power of speech, if you could
call theirs a power. The Fuegian vocabulary was so small,
and their grunt-sunken grammar was so simple and simpleminded, it was a rather lame distinction, to Darwin's way of
thinking.[22] He had no idea yet that speech, whether grunted
by brutes in the middle of nowhere or intoned by toffs in
London, was by far—very far—the greatest power possessed by any creature on earth.

It was after laying eyes on these and other hairy apes
below the equator that a blasphemous, mortally sinful, absolutely exciting, fame-flirting, glory-glistening notion stole
into Darwin's head. What if people like the Fuegians
weren't really *people* but rather an intermediate stage in
the transmutation, the evolution, of the ape into…*Homo
sapiens?* That God created man in his own image was a
centerpiece of Christian belief. In 1809, when Lamarck had
dared to suggest (in *Philosophie Zoologique*) that apes had
evolved into man, it was widely assumed that only his leg-

endary heroism during the Seven Years' War saved him from serious grief at the hands of the Church and its powerful allies. (Artillery fire had killed more than half of a French infantry company's men and all its officers. A short, skinny, seventeen-year-old enlisted man, a Private Lamarck, stepped forward and through sheer force of personality assumed command and held the company's position until reinforcements arrived...)

Darwin was petrified by the prospect of condemnation but aflame with ambition. Seven years later, in 1844, the author of *Vestiges of the Natural History of Creation* felt the same way: so he hid behind the byline Anonymous and never came out. Not even the prospect of fame was enough to overcome his fear. His name was not revealed until the twelfth edition of *Vestiges* was published in 1884, the book's fortieth anniversary...thirteen years after the author's death. Then, at last, the title page bore a byline: Robert Chambers. For all their snobbery, the Gentlemen naturalists proved to have been right. Chambers was not a Gentleman but a journalist, cofounder with his brother, William, of *Chambers's Edinburgh Journal* and *Chambers's Encyclopaedia*...and an amateur one-shot naturalist.

Darwin turned out to be every bit as gun-shy as Chambers, but more than that held him back. As a dedicated naturalist he had an even bigger problem: a huge gap in evidence when it came to language, which set humans far apart from any animal ancestors. It gnawed at him. He

19

could explain man's opposable thumb, upright stature, and huge cranium, but he couldn't find one shred of solid evidence that human speech had evolved from animals. Speech seemed to have just popped up into the mouths of human beings from out of nowhere. He thought and thought. And thought...

Wait a minute. What *was* speech? Vocal communication, right? Well, animals had their forms of vocal communication, too, and some were fairly complex. Vervet monkeys had different cries to warn the troop of the presence of the most dangerous predators. They had one cry for leopards, another for eagles, another for baboons, another for pythons, plus variations of the python cry to indicate a mamba or a cobra. They used certain intonations to indicate that a particular fellow vervet's reports weren't altogether reliable. There you had it: monkey semantics. If that wasn't the equivalent of speech, what was? All right, there was no direct evidence to point to...but it was *self*-evident, wasn't it? Animal speech like the vervet's had evolved into human speech...somehow...and if there was no clear evidence...well, it just meant nobody had looked hard enough, because it *had* to be there somewhere.

But why *had to be?* Because at that moment, in 1837, Darwin had fallen, without realizing it, into the trap of cosmogonism, the compulsion to find the ever-elusive Theory of Everything, an idea or narrative that reveals everything in the world to be part of a single and suddenly clear pattern.

The first savant to set such a goal seems to have lived during the third century BCE, although the term itself, Theory of Everything, wasn't coined until half a century ago by a science fiction writer, Stanislaw Lem, in order to make fun of it. By the end of the last century, it had started appearing in scientific journals with a serious face on. To Darwin it had been serious business a century before that, no matter what it was called. Proving that speech evolved from sounds uttered by lower animals became Darwin's obsession. After all, his *was* a Theory of Everything. No matter what verbal acrobatics and leaps of logic it might require, speech, language, *had to* fit into his flawless cosmogony. *Speak,* beasts.

"Cosmogony" literally means "world-birth." In its pure form, a cosmogony is an account, like the one in the Bible's book of Genesis, of the creation of the universe and all forms of life, culminating in man. In the beginning, nothing material exists, only a spirit and a force called God. God creates the material world in six days and rests on the seventh. He creates man in his own image and puts him in charge. Very few other cosmogonies feature any such almighty god or great invisible force as the creator. Quite the opposite.

The vast majority of cosmogonies involve an animal, and the animal is never one noted for its size, physical power, or ferocity. Not at all. The trend is not toward bigger and bigger but smaller and smaller. One version of the North American Apache cosmogony opens with a great void. Way up in the void arrives a disk. Curled up inside the disk is a little old man with a long white beard. He sticks his head out and finds himself utterly alone. So he creates another little man, much like himself. (Kindly refrain from mundane technical questions.) Somehow, up in the void, they take to playing with a ball of dirt. A scorpion appears from nowhere and starts pulling at it. He pulls whole strands of dirt out of the ball. Longer and longer he pulls them, *farther farther farther* they extend, until he has created earth, sun, moon, and all the stars.[23] This is, of course, the original version of the current solemnly accepted—i.e., "scientific"—big bang theory, which with a straight face tells us how something, i.e., the whole world, was created out of nothing. What this, like virtually every other contemporary retread of an ancient cosmogony, lacks is the original's cast of colorful characters. The big bang theory desperately needs someone like the scorpion or the little man with a long white beard curled up inside a disk. Or like Michabo the Great Hare. Michabo is the creator in the Algonquin Indians' cosmogony. In the beginning, the Algonquins believed, the earth's surface was entirely under water. There was no visible land at all. One day Michabo the Great Hare is out on a raft with a crew of other animals. He tells three of them to dive to the bottom of the sea and bring up some earth. They manage to find a single grain of sand...out of which the Great Hare creates a huge island, apparently

North America. He transmutes the bodies of dead animals into men, namely, the Algonquins.*

Compared to the animals who star as creators in other cosmogonies, however, a full-grown hare is an enormous creature. Many are mere birds. The Tlingit natives of the American Northwest believed they originated with Yehlh the Raven. Yehlh creates the world and populates it with man—but there's no *light!* Everything is pitch dark. You can't see your hand in front of your face. The problem is, Yehlh has a Good-versus-Evil, God-versus-Devil relationship with a dark-feathered uncle who has stolen the sun, the moon, and the stars. So Yehlh turns himself into a hemlock needle that gestates into a boy. Just a boy, he looks like, and the uncle thinks nothing of it. The boy discovers the sun, the moon, and the stars hidden in a box, runs away with them, then turns himself back into Yehlh the Raven and flies them to the heavens and lights up the world.[24] The Cherokee Indians' cosmogony resembled the Algonquins' but starred an animal that makes a raven look gigantic, namely, a water beetle. It is Beetle who dives to the bottom of the primal sea and comes up carrying the bottommost mud, which he will use to create the earth.[25] Yet another so-called "earth diver" cosmogony, the Assiniboine Indians', stars an insect smaller than Beetle, namely,

---

* Michabo is a prominent figure in Algonquian folklore. The Powhatan Museum, in Washington, DC, provides more information on his role and mythology.

Inktomi the Spider. Inktomi makes the dives, creates earth, and populates it with human beings...plus horses for them to ride.[26] In an Egyptian cosmogony a dung beetle called Khepri takes on the persona of Atum-Re, god of the morning sun. He resurrects himself every morning, rising from the underworld. He could *use* a new persona. Dung beetles live by rolling other animals' offal into balls and eating them or hiding them in the ground for later. The Egyptians gave the dung beetle a name that wasn't exactly music to the ears, but you could at least say it in front of company, viz., the scarab.[27] Among the Khoisan peoples in Africa, the great cosmogonic creator is Cagn, a praying mantis, an insect that looks positively anorexic next to a svelte, fat dung beetle. Cagn created not only all animals and all people but also language...and the moon. The moon was an afterthought. One night some hunters kill an animal Cagn has created. So Cagn takes its gallbladder out and hurls it into the sky and lights it up...to give animals—and people—a fighting chance of seeing what's coming at night.[28]

The Navajo Indians' creator was an insect that seems to have been identical to what is known today as the biting midge (colloquially, the no-see-um bug). Biting midges are so small you can't see them. But you can't mistake them when they bite your ankle. For all practical purposes they are invisible. But the Navajo biting-midge creator was smaller than small and invisibler than invisible, because it came into this world without its two wings. Yet it is the creator in probably the most sophisticated cosmogony ever believed in, a story of full-scale, gradual evolution from next to nothing to modern man. In the beginning, the biting midges lived in the First World, down deep deep deep beneath the earth's surface. As evolution began, they grew back their missing wings, and one species evolved all the way into a full-blown insect, a locust. Locust led the hives up into the Second World, where they began to evolve into animals of every species. Then Locust led the whole burgeoning menagerie up into the Third World, where the most advanced species evolved into men. Then Locust led all the men and all the animals up into the Fourth World, which was right below the crust of the earth. In an Inktomi-like show of energy and dedication, the menagerie's spiders built rope ladders out of their webbing so that everybody could climb up onto the earth's surface.*

A later cosmogony was a dead ringer for the Navajos', dead and unfortunately duller, except for one thing. The creator in this cosmogony was a creature even smaller, even less visible to the naked eye, than a biting midge, namely, a single, undifferentiated cell—or "four or five" of them. "Undifferentiated" means it could evolve into any living thing, vegetable or animal. This cosmogony was the only one recent enough for people to know the chief storyteller

---

* The Utah Department of Heritage & Arts and the Navajo Nation's websites contain retellings very similar to this version.

by name: Charles Darwin. "Four or five" is from a scrap of conversation he had with a group of students not long after he told the story publicly. The students had the sort of naive, unbridled, free-floating curiosity most youths unfortunately rein in far too early in life. They wanted to know some small but fundamental details about the moment Evolution got under way and how exactly, physically, it started up—and from what?

Darwin had apparently never thought of it quite that way before. Long pause...and finally, "Ohhh," he said, "probably from four or five cells floating in a warm pool somewhere."* One student pressed him further. He wanted to know where the cells came from. Who or what put them in the pool? An exasperated Darwin said, in effect, "Well, *I* don't know...look, isn't it enough that I've brought you man and all the animals and plants in the world?"

In this respect, Darwinism was typical of the more primitive cosmogonies. They avoided the question of how the world developed ex nihilo. Darwin often thought about it, but it made his head hurt. The world was just...*here*. All cosmogonies, whether the Apaches' or Charles Darwin's, faced the same problem. They were histories or, better said, stories of things that had occurred in a primordial past, long

---

* Darwin used similar language in a letter to J. D. Hooker dated February 1, 1871, and also in *The Origin of Species*.

before there existed anyone capable of recording them. The Apaches' scorpion and Darwin's cells in that warm pool somewhere were by definition educated guesses. Darwin, a Cambridge man, after all, was highly educated by the standards of his time, but so, no doubt, was the Apache medicine man who came up with the little old man with the long beard in the disk. The difference in Darwin's case was that he put together his story in an increasingly rational age. It wouldn't have occurred to him to present his cosmogony as anything other than a scientific hypothesis. In the Navajo cosmogony the agent of change (as distinct from the creator) was alive. It was Locust. In Darwin's cosmogony it had to be scientifically inanimate. Locust was renamed Evolution.

There were five standard tests for a scientific hypothesis. Had anyone observed the phenomenon—in this case, Evolution—as it occurred and recorded it? Could other scientists replicate it? Could any of them come up with a set of facts that, if true, would contradict the theory (Karl Popper's "falsifiability" test)? Could scientists make predictions based on it? Did it illuminate hitherto unknown or baffling areas of science? In the case of Evolution...well...no... no...no...no...and no.

In other words, there *was* no scientific way to test it. Like every other cosmogony, it was a serious and sincere story meant to satisfy man's endless curiosity about where he came from and how he came to be so different from the animals around him. But it was still a story. It was not evidence. In short, it was sincere, but sheer, literature.

It certainly wasn't scientific experimentation or observation that finally convinced Darwin that man had no special place in the universe. It was a visit to the London Zoo in the spring of 1838, two years after the voyage of the *Beagle*. One of the zoo's most popular attractions was an orangutan named Jenny. Jenny had become so used to being around people that many of her reactions had become absolutely human in nature. Sometimes she wore clothes. Her gestures, her facial expressions, the sounds she made, the way she acted out frustration, mockery, anger, guileless glee, or *I-love-you, Help-me! Help-me!, I-want I-want*...this last with a whine that made one see how hard she was struggling to put it all into words—it was clear as day! Now Darwin was certain! Jenny was a human being behind the flimsiest of veils. He used his clout as a Gentleman and a leading naturalist to enter Jenny's cage and study her expressions up close.[29]

Certain he was...and so what? That left him as stumped as everybody else who was so sure about it, including his grandfather Erasmus Darwin. Erasmus couldn't figure out exactly *how* transmutation*—Evolution—occurred, and neither could his grandson.

---

* Darwin preferred the term "transmutation."

In October of 1838, Charles happened to pick up a copy of Thomas Malthus's *Essay on the Principle of Population*... "for amusement," as he put it, apparently assuming that no deep thinker could possibly find a book as old and popular as *Principle of Population* profound.[30] He started reading it, and—

Ahura! That old Malthusian magic's got me in its spell! It lights up Darwin's brainpan precisely the way it would Wallace's twenty years later—*It!*—the solution to what naturalists, including Darwin himself until that very moment, called "the mystery of mysteries": how the littlest creature (or "four or five" of them), smaller even than the smallest invisible biting midge—namely, a cell; never mind those great bulky hares and scorpions and dung-eating beetles—a *cell,* or a cell and a few brethren, grew up into the most highly developed creature of all, one with a certified Latin name, *Homo sapiens.*

But what happens to someone like Darwin, who has been honored, who is highly esteemed, who has the highest credentials in his field...when he announces that man is not made in the image of God but is, in fact, nothing but an animal? He could see, *feel,* the Church and thousands, tens of thousands, of believers descending upon...*me*...with the Wrath of God. He was aware of what had happened to Mr. *Vestiges,* all too aware. It terrified him.

So in the two decades between 1838—back when he

was twenty-nine years old—and 1858 he hadn't told a soul other than Lyell about his *Ahura!* moment, and he didn't even tell Lyell until 1856. For the twenty years before that, his career had been devoted...secretly...to compiling evidence to support what in due course, he calculated, would shake the world: his revelation of the *actual* origins of man—and, while he was at it, all animals and plants: his Theory of Evolution through natural selection. He was bringing forth, for all mankind to marvel at...the *true* story of creation! Man was not created in the image of God, as the Church taught. Man was an animal, descended straight from other animals, most notably the orangutans.

Darwin was afraid of not one but two things: one, the Wrath of the Godly, and two, some enterprising competitor getting wind of his idea and forestalling him by writing it up himself. And sure enough, up from nowhere pops this little flycatcher Wallace with a scholarly paper, ready for publication, on the *evolution of species through natural selection!* He racked his brain to recall whether or not he had written something in a letter that tipped Wallace off. But he couldn't recall a thing.

Oh, Lyell had warned him*...Lyell had warned him... and now all my work, all my dreams—*all my dreams*—

Then he caught hold of himself. He mustn't give in to

---

* Lyell had encouraged Darwin to publish his ideas on evolution before someone beat him to it. This is clear in a letter Darwin wrote to Lyell on June 18, 1858, after receiving Wallace's paper.

this horrible feeling overwhelming his solar plexus. There was something more important than priority and glory and applause and universal admiration and an awesome place in history...namely, his honor as a Gentleman and a scholar. He summoned up every tensor of his soul and did what he had to do, and he did it like a man. He dispatched Wallace's paper to Lyell along with a letter saying, "It seems to me well worth reading...Please return to me the M.S. which he does not say he wishes me to publish; but I shall of course at once write & offer to send it to any Journal. So all my originality, whatever it may amount to, will be smashed."*

Coming from the pen of a Gentleman as ever-composed and self-possessed-to-the-point-of-phlegmatic as Darwin, that word "smashed" rose up from the page like a howl, a howl plus the *riiiippp* of those tensors in his soul going haywire and tearing the damned thing to pieces. What he howled was, "My *whole life* is about to be *smashed* and reduced to dust, to a mere footnote to the triumph of another man!"

* From the letter of June 18, 1858.

CHAPTER II

GENTLEMEN AND OLD PALS

OH, CHARLIE, CHARLIE, CHARLIE...said Lyell, shaking his head. Who was it who warned you two years ago about this fellow Wallace? Who was it who told you you'd better get busy and publish this pet theory of yours?...So why should I even bother, this late in the game?

But...we *are* Gentlemen and old pals, after all...and I think I know of a way to get you out of this predicament. It so happens there is a meeting of the Linnean Society, postponed from last month in deference to the death of one of our beloved former Linnean presidents, coming up *thirteen days from now,* July 1. Unfortunately, we don't have any way to notify Wallace in time, do we. But that's not *our* fault. *We* didn't schedule the meeting. That's just the way it goes sometimes. We'll bring our good friend Sir Joseph Dalton Hooker, the botanist, in on this. All three of us are on the society's council.* We can make the whole thing seem like the most routine scholarly meeting in the world...the usual learnéd papers learnéd papers learnéd papers, the usual *drone drone drone humm drumm humm*...The main thing, Charlie, is to establish your priority. We'll present your work *and* Wallace's. Now, that's fair, isn't it? Even-steven and all that? Well, to be perfectly frank, there *is* one slight hitch. You've never published a line of your work on Evolution. Not one line. As far as the scientific world at large is aware, you have never *done* any. You don't even have a paper to present at the meeting...*hmmm*...Ahh! I know! We can help you create an abstract overnight! An *abstract*. Get it?

"Abstract" was the conventional word in scientific publications for a summary of an article. It usually ran right below the title. After that came the article itself.

Now do you get it, Charlie? All we need is for you to give us an *abstract* of a scholarly paper of yours that doesn't exist!

Darwin was aghast. "I should be *extremely* glad *now* to publish a sketch of my general views in about a dozen pages or so," he wrote to Lyell. "But I cannot persuade myself that I can do so honourably...I would far rather burn my whole

---

* See the Darwin Correspondence Project database for the letter sent by Hooker and Lyell to J. J. Bennett, Esq., the secretary of the Linnean Society. On June 30, 1858, they wrote to request that Darwin's and Wallace's papers be presented at the following day's meeting (www.darwinproject.ac.uk/entry-2299).

book than that he or any man should think that I had be-
haved in a paltry spirit."*

In fact, he said, he had been intending to write Wallace
relinquishing all claim of priority when Lyell's letter arrived.
So how could he possibly concoct his own essay overnight
and raise his hand and claim priority himself? How very
paltry. Darwin had taken to repeating this word, "paltry,"
over the last few days. It meant "small-minded," "mean,"
"vile," "despicable." Not a pretty word, but a lot better
than "dishonest"—

—hold on a second: what's *that?* A tiny hole...or is he
just seeing things?...no, there's a tiny hole in Lyell's letter,
the tiniest hole you ever saw in your life...and through that
hole shines a little gleam of light, so little he wonders if it
could possibly be real...but it *is* real! It emits the faintest
of glimmers—the faintest of glimmers, but an *honourable*
glimmer!

Darwin pivots 180 degrees. His heart turns clear around.

Or that "was my first impression," he says to Lyell, suddenly switching gears, "& I should have certainly acted on
it, had it not been for your letter."[1] But your *letter...your
letter...your letter* has shown the way. *Even Stephen,* you
have ruled, *Even Stephen!* And who am I to presume to
overrule Sir Charles Lyell? You are the dean of British nat-

---

* Taken from a letter Darwin wrote to Lyell on June 25, 1858. For
  the complete letter, see the Darwin Correspondence Project database
  (www.darwinproject.ac.uk/entry-2294).

uralists, my old friend. There is no greater or wiser man in this entire field. Everyone, including Wallace, will be better off in the end if we leave all this in your accomplished hands.

Yes, Sir Charles Lyell had made his decision. The two papers, Wallace's and Darwin's, would be made public simultaneously before the Linnean Society. With a single stroke Sir Charles had made the question of priority disappear. He, Darwin, would not be claiming priority. Just the opposite. He was extending a magnanimous hand to a newcomer. He would be making room on the stage for a lowly flycatcher to be heard.

The one remaining catch was that Lyell and Hooker expected Darwin to write his own abstract. He couldn't do that—he *mustn't* do that. He begged off with some pathetic excuse. He didn't have the courage to tell them that his own conscience must be kept clear. His *own conscience* had to believe he had nothing to do with this project. It wasn't *his* idea. It was entirely theirs, Lyell's and Hooker's. I, Charles Darwin, *had nothing to do with it!* Above all, let no man be able to say I wrote an abstract for myself after reading Wallace's paper. There mustn't be a hint of any such paltriness before an august body like the Linnean.

So it fell upon Lyell's and Hooker's shoulders, the task of concocting for Charlie an abstract out of what they could lay their hands on quickly... let's see... we have a copy of a letter he sent last year to an American botanist at Harvard named Asa Gray giving a halfway outline of his concept of natural selection…and there's some sort of abortive "sketch," as he calls it, that he has at home for a book on transmutation he's been telling himself…for the past fourteen or fifteen years…he's going to write someday.* And of course we have Wallace's paper for…*hmmmm*…how should one put it?…for "background" or "context" or maybe something along the lines of "corroborative research" or "heuristic monitoring." We'll think of a term. In any case, we're in a position to make sure there will be no important points in Wallace's paper that aren't also in Charlie's. Hooker's wife, Frances, is a bright little number. We'll get her to read Wallace's letter and then pull together some extracts from Charlie's "sketch"…and, while she's at it, shape things up a bit…where necessary.** There is more than one way to swat a flycatcher.

When they were finished, Darwin had *two* papers to his

---

* Darwin sent Hooker copies of the letter to Asa Gray, an 1844 sketch of his theory (which Hooker had previously annotated), and Wallace's paper. These items were accompanied by one of two letters sent on June 29, 1858, in response to Hooker's request. In both June 29 letters, Darwin alludes to the fact that Hooker and Lyell will be drafting a manuscript for submission to the Linnean Society. For the complete text of these letters, see the Darwin Correspondence Project database (www.darwinproject.ac.uk/entry-2297 and www.darwinproject.ac.uk/entry-2298).

** Darwin thanks Frances Hooker in a letter to her husband dated July 5, 1858 (available at the Darwin Correspondence Project database, www.darwinproject.ac.uk/entry-2303), and again on July 13, 1858 (www.darwinproject.ac.uk/entry-2306).

name, both very short—first, an abstract of his letter to Asa Gray and, second, the extract of his unpublished sketch, tidied up by Mrs. Hooker. Combined, they were almost as long as Wallace's twenty pages.

To put the matter in perspective, one has only to imagine what would have happened had the roles been reversed. Suppose Darwin is the one who has just written a formal twenty-page scientific treatise for publication...and somehow Wallace gets his hands on it ahead of time...and announces that he made this same astounding epochal discovery twenty-one years ago but just never got around to writing it up and claiming priority...a horse laugh? He wouldn't have rated anything that hearty. Maybe a single halfway-curled upper lip, if anybody deigned to notice at all.

At the Linnean Society meeting on July 1, neither party was present...not Wallace, because the Gentlemen had been more than content to leave the flycatcher in the dark in equatorial Asia, 7,200 miles away...and not Darwin, because his infant son, Charles Waring Darwin, his and Emma's tenth child after nearly twenty years of marriage, had died of scarlet fever on June 28.* He couldn't very well show up in public three days later, on July 1, advancing his career beneath a banner saying HUMAN BEINGS ARE NOTHING BUT ANIMALS.

* From the letter of July 5, 1858.

At Linnean Society meetings, papers on a single subject were read in alphabetical order, by author, and—wouldn't you know it?—*D* comes before *W*.[2] That was just the way it goes sometimes, too. So the society heard two of its most distinguished members, Sir Charles Lyell and Sir Joseph Dalton Hooker, peers of the realm, do the introductions, which were spent pointing out that Darwin, who clearly had priority, was all for including Wallace on the program.* Both authors may "fairly claim the merit of being original thinkers in this important line of inquiry," Lyell and Hooker begin, but Darwin was the *first*...it just took him twenty-one years to get around to writing his thoughts down. Then the Linneans heard not one but two papers by their renowned colleague Charles Darwin, member of the Royal Society of London, famous for his many years of worldwide explorations...and then one by some little flycatcher named Wallace. It was not hard to get the impression that the distinguished Mr. Charles Darwin, with his big heart, was giving a pat on the head to this obscure but promising young man off catching flies in the tropical bowels of Asia.

That impression never changed. Wallace was an outsider and not a Gentleman, not the Linnean Society sort. An undersecretary read the introduction and all three papers

* The papers were introduced by a reading of the June 30, 1858, letter from Lyell and Hooker to the secretary of the Linnean Society. See also Charles Darwin and Alfred R. Wallace, "Proceedings of the Meeting of the Linnean Society held on July 1st, 1858," *Journal of the Proceedings of the Linnean Society: Zoology* 3.

aloud. They prompted no questions or discussion; none at all. Most of the twenty-five or thirty Linneans on hand appeared bored, if not put to sleep, by the drizzle drizzle of species transmutation biogeographical variations injurious adaptations drizzle drizzle...when, O Lord, will the fog clear out? They had come to hear Lyell, a gentleman among Gentlemen, deliver a promised eulogy of the society's lately departed former president, which he did, first thing. As for the rest of the program, they did their best to endure...the first public revelation of a doctrine that would turn the study of man upside down—and kill God, if Nietzsche had anything to say about it. At the moment, however, the Gentlemen of the Linnean Society greeted the news with yawns so big they couldn't cover them with their bare hands.

In his annual state-of-the-society speech the following spring, the society's president said, "The year which has passed...has not, indeed, been marked by any of those striking discoveries which at once revolutionize, so to speak, the department of science on which they bear."[3]

It was not until three months later, in October of 1858, that Wallace, by then off to New Guinea for more flycatching, had any idea that a meeting of the Linnean Society involving his work had ever taken place. The news arrived in letters (both in the same envelope) from Hooker and Darwin, implying how generous Darwin had been throughout and how highly he thought of Wallace.* He had given him equal credit not only before the Linnean Society but also in an upcoming issue of the *Journal of the Proceedings of the Linnean Society*. Wallace's paper and Darwin's "abstracts" would be running even-steven in its prestigious pages. Once it was published, Darwin got the courage to send Wallace a copy. The truth was, the very layout and look of the *Journal* were so unbalanced in Darwin's favor that Darwin himself sucked in a great guilty two-whole-lobes load of air and averted his eyes the moment he saw it. The distinguished Mr. Darwin's name was placed first on the contents page and at the top of page after page after that—the strictures of alphabetical order again,[4] wouldn't you know it—and the unknown Mr. Wallace's followed...another generous pat on the head for an obscure young man who certainly had worked-hard-you-had-to-hand-him-that.

Wallace hadn't a clue that his paper was going to be published. He had sent it to Lyell for an expert opinion before publication. After all, he would be presenting the world with a radical discovery: how Evolution occurred through natural selection. He hadn't the faintest notion that

* The original letters are lost, but Wallace's response to Hooker (dated October 6, 1858) provides some information on their contents. (See the Darwin Correspondence Project database, www.darwinproject .ac.uk/entry-2337.)

the Linnean Society—meaning three of its officers, Darwin, Lyell, and Hooker—had their hands on it and would do with it as they pleased. They hadn't asked for permission and never gave him a chance to edit or proofread his own work. They would never have dared pull any such sleight of hand on a Gentleman—*any* gentleman, no matter how obviously clueless he might be.

Wallace's reply, a letter to Hooker, got straight to the sore point. "I was very much surprised to find that the same idea had occurred to Darwin."[5] But then he gave up. He knew nothing at this juncture about exactly how it had "occurred" to Darwin, and he didn't have the nerve to insist on finding out. He realized there was no way that he, all by himself on the wrong side of the class divide, was going to prevail against the Gentlemen. He was a flycatcher. He might as well be content to keep his mouth shut and salvage what he could from the wreckage of his plans and let them dress him up in their flattery and pluck him up from obscurity and put him on the Big Stage. He couldn't have sounded more grateful. By the time Wallace got the letters from Darwin and Hooker, Darwin had already been writing for three months, faster than he had ever written in his life, to scoop Wallace by publishing that most solid, hard-shelled claim to priority: a book.

Three months' head start—but where was he ever going to find the energy to finish the race? Throughout the voyage of the *Beagle*...way back when...Darwin had been in his twenties, enjoying the heedless animal health of youth. To-day, in October of 1858, going on twenty-two years later, he was almost fifty... and afflicted with what his doctors told him was dyspepsia. But very likely their real diagnosis was hypochondria... referring to some recurring imaginary malady unlikely to kill you even if it were real. In Darwin's case it consisted of sudden, uncontrollable vomiting and every sort of pain in his distended belly and bowels, every known belch, retch, heave, gas-pass, watery rush, and loathsome gush, plus foul wind erupting from one end of his digestive tract and foul sounds eructing *grrrrekkk* from the other. And where was he going to find the time? Half the time he seemed to be laid up in the Ilkley spa, in the Yorkshire Dales, taking "the waters" and "the cure," wrapped up in wet sheets from head to toe like a mummy in order to douse the fiery itch of his chronic eczema.*

Only his atrial-fluttering fear of Wallace somehow *smashing* his *whole life* by producing a book of his own, once again forestalling *me* and establishing *my* own *priority,* this time beyond the reach of any more monkey business by my Gentleman sidekicks—only this kept him out of Ilkley's sopping cemetery long enough to write his hard-shelled

---

* Darwin's health problems have been the subject of much speculation. He discussed his symptoms and treatment in his personal correspondence, and Emma Darwin kept notes on his health and sleeping habits in her diary. Biographer Janet Browne takes a more contemporary approach in *Darwin's Origin of Species* (New York: Atlantic Monthly Press, 2007).

claim to priority. The flycatcher was still in Malay, so far as Darwin knew. But that hadn't kept him from being the actual creator of the Theory of Evolution the first time around. So who knew what he was concocting right now?

Darwin gave his book a twenty-one-word academic title, but common usage would quickly pare it down to four: *The Origin of Species*. Late September of 1859...and Darwin was going over the last details for publication, set to take place in two months, which would be late November—and no sign of Wallace anywhere, so far. He began to let his breath out slowly and slowly let his hopes rise. But he remained at Ilkley still, wrapped up in the spa's wet mummy sheets, still enduring a raging case of eczema and still trembling...for a reason that had nothing to do with Wallace. He hadn't dared push his theory all the way to its shocking conclusion, which would be the news, the revelation, that man did not come into this world in the image of God but out of the loins of an orangutan or some other big ape. Man was an animal and nothing but an animal. If he took it that far, all the way at once...he shuddered to think of how violent the reaction would be—the rage! the fury!—from the Church and the clueless Christian middle classes. He could see all his honors and medals and elite memberships crashing to earth amid the ruins of the reputation he had so single-mindedly aspired to ever since the *Beagle* returned home twenty-two years ago. So in *The Origin of Species* he drove the Theory of Evolution right up to *Homo sapiens*'s front door but not one inch closer...unless you counted a single, soft, one-knuckle tap two pages from the end of the book, offering a cryptic hint as to where he might be heading in a sequel, if he should he ever write one.

"In the distant future I see open fields for far more important researches. Psychology will be securely based on a new foundation, that of the necessary acquirement of each mental power and capacity of gradation. Light will be thrown on the origin of man and his history."[6]

One faint cryptic hint too many, old chap! On November 19, five days before publication, an anonymous reviewer in the prestigious journal *Athenaeum* eviscerated the book and fried the entrails. A single sentence in the piece leaped out at Darwin: "If a monkey has become a man—what may not a man become?" *A man!* says his nameless assailant! At the time virtually all book reviews were unsigned, the theory being that anonymity gave the reviewer the freedom to be frank. But it wasn't supposed to give him carte blanche to make vicious distortions! This one had gone straight to that short obiter dictum two pages from the end and made it seem like the whole book is about man thrashing and splashing and gibbering away in some primordial muddy puddle somewhere. The message was: Don't risk your sanity trying to read it! Leave that to the philosophers and divines who enjoy dog-paddling around in such slop.

"If a monkey has become a man!"...and Darwin thought he had so cleverly kept man hidden in the wings...So much

for that delusion. Right away this bastard spots man peeking out from behind a curtain. *This bastard*—Darwin was never one to resort to off-color language, but then no one had ever hurt and humiliated him and dashed his hopes this thoroughly, either. *The first review!*

On November 21, 1859, Hooker writes Darwin to say that he and Lyell think the asinine anonymous *Athenaeum* reviewer must be a geologist named Samuel Pickworth Woodward. Forever after, Woodward, hopelessly baffled, flinched whenever he found himself in the same room with Darwin. Darwin cut him dead every time or else gave him an iceberg. One-eighth of an iceberg's mass sticks up above the frigid surface with a tip of thirty-three-degree civility. The other seven-eighths is hidden under water...a gigantic ice boulder of frozen loathing and resentment, hard as a rock. In fact, the nameless hugger-mugger was another naturalist entirely, an Anglican priest named John Leifchild.

The *Athenaeum* blast so tenderized Darwin that he failed to understand what was happening when a regular storm of reviews and commentaries erupted during December and the first six months of 1860. Even mildly negative reviews hit him like body blows. The fierce ones cut him clear through to the gizzard. The *Edinburgh Review* ridiculed not only his theory but also his prose style, his scientific ignorance, his scholarly incompetence, all of it lazily afloat in his shallow brain. One had only to compare Darwin, the piece went on, with someone like, say, Britain's preeminent naturalist and president of the British Association for the Advancement of Science, Richard Owen.[7] Now, *there* we're talking about a deep thinker, a real scientist: Richard Owen, Richard Owen, Richard Owen. Owen's name kept coming up. Darwin went over the review repeatedly. He couldn't believe what he was looking at: as usual, it was unsigned...but the man's finicky rhetoric and would-be-cosmopolitan displays of how much French he knew gave the game away. It was his longtime friend-he-had-always-assumed...Richard Owen. He never spoke to Owen again.

By then he was in such a wary, defensive state of mind that even the positive reviews struck him as tepid or tentative—with one exception: an absolute rave in the very voice of the British upper orders, the London *Times*. The *Times* ran only one or two book reviews per month. Like the others, this review was published anonymously. But Darwin soon learned it was written by one of his younger adherents, the anatomist Thomas Huxley.[8] In a piece of dumb luck, Huxley had happened to run into the writer the *Times* had assigned to do the review. The man was moaning that he wasn't even remotely familiar with the subject. Huxley came up with the bright idea of writing the piece for him—anonymously, needless to say. Darwin wound up with an astounding boost in the mighty *Times*. Huxley became the best public relations wizard any scientist had ever had.

Huxley's background was similar to Alfred Wallace's, although their personalities could scarcely have been more different. Huxley's father was an up-against-it mathematics teacher who couldn't pay for his son's education beyond two years of grade school. The boy became a scientific prodigy all the same, a largely self-taught anatomist. At nineteen he discovered an internal component of hair no biologist had ever dreamed existed. By age twenty he had the pleasure of seeing it referred to in scientific journals as "Huxley's sheath." It was the first of a series of anatomical discoveries he would make. He was only twenty-five when he was elected a member of the Royal Society.

The boy wonder was such a hot number in scientific circles that Darwin courted him as an acolyte, and the boy came through for him in a big way. He wrote five long, enthusiastic reviews of *The Origin of Species* in major journals in the space of four months, the two longest conveniently anonymous, and that was the least of it.* In person he was a good-enough-looking man, but with a bulldog's build, a bulldog's neck, and a bulldog's prognathous jaw when he was angry, which was often, since he loved a good fight. He was aware of all that and enjoyed being called "Darwin's bulldog." In June of 1860, he starred in a much-written-about British Association for the Advancement of

---

* Huxley published a second anonymous review in the *Westminster Review* (April 1860). He also published credited reviews in *Macmillan's Magazine* (April 1860), *The Medical Circular* (March 7, 1860), and the *Proceedings of the Royal Institution of Great Britain* (February 10, 1860).

Science debate over Evolution against the Church of England's most renowned public speaker, Bishop Samuel Wilberforce. He went on to create the X Club, a group of nine prominent naturalists, including Hooker, who met every month at some restaurant or a club and set about—very successfully—stacking influential university science faculties with Darwinists. The X Clubbers had a big hand in creating the pro-Darwinist journal *Nature* (which thrives to this day).[9] They attacked every Darwin doubter the moment he dared raise his voice. That mode of intimidation only intensified over time, leading to what is still known today as "the Neo-Darwinist Inquisition."[10]

Huxley became such an ardent Darwinist not because he believed in Darwin's theory of natural selection—he never did—but because Darwin was obviously an atheist, just as he was. No one dared flaunt such a loaded term, of course. Huxley said he was not an atheist but an agnostic. He made up the word. An agnostic, he said, was the opposite of a *gnostic*.[11] Gnostics held an early Christian and even pre-Christian belief that people should separate knowledge of the material world from the only true knowledge: the spiritual. An *agnostic* like him wasn't even sure there was a God. This newest Huxleyism entered the language the way "Huxley's sheath" had.

Huxley's great PR campaign happened to coincide with two sweeping mid-nineteenth-century developments in western Europe—Britain, especially—creating, as the phrase goes, a perfect storm. One was the sudden proliferation of magazines and newspapers, whipping up a competition not only for hard news but also for stories of every sort of social and intellectual trend…such as the Theory of Evolution. The second was what the German sociologist Max Weber called "the disenchantment of the world." Well-educated, would-be-sophisticated people all over Europe had begun to reject the magical, miraculous, superstitious, logically implausible doctrines of religion, such as the Virgin Birth of Christ, the Creation (of the world in seven days), Christ's Resurrection, the power of prayer, the omnipotence of God, and a thousand other notions that were irrational by their very nature. Three decades earlier, Coleridge had concluded that the influence of the clergy was fading so rapidly that he revived the by-now-obsolete term "clerisy." The clerisy, he said, were the secular thinkers who had replaced the clergy here in the nineteenth century…in matters spiritual as well as philosophical.* Near the end of the century, while the Dreyfus case raged in France, the country's off-and-on president, Clemenceau, would call them (with a nod toward Anatole France and Émile Zola) "the intellectuals," and that was the name that stuck, in England as well as France.

The Theory of Evolution eliminated all such mystification. At the higher altitudes of society, as well as in

---

* Coleridge first used the term in *On the Constitution of the Church and State According to the Idea of Each* (1830).

academia, people began to judge one another socially according to their belief, or not, in Darwin's great discovery. Practically all Church of England clergymen were well educated and well connected socially, and by 1859 the demystification of the world had extinguished whatever fire and brimstone they might have had left. The sheerly social lure of the theory, the status urge to be fashionable, was too much for them. Subscribing to Darwinism showed that one was part of a bright, enlightened minority who shone far above the mooing herd down below. There were plenty of clerical attacks on *The Origin of Species,* but they were so civil and rhetorically well mannered that the new agnostics didn't cringe in fear of an angry God, much less a vengeful one. The theory and the atheistic bias that came with it spread quickly to Germany, Italy, Spain, and to self-professed intellectual elites in the United States, even though the great mass of the population kept on mooing and made sure America remained the most religious country on earth outside of the nations of Islam (and it remains so today).

Only in France was Darwin written off as just another little man with a big theory. It took three years for *The Origin of Species* to find a French publisher. France had gone through its own Evolution debate—the French term was "transformism"—thirty years earlier, mainly thanks to Lamarck's influence. But the leading French spokesman for transformism, Étienne Geoffroy Saint-Hilaire, made the mistake of taking on Georges Cuvier in debate. Cuvier, a zoologist, paleontologist, anatomist, politician, and aristocrat, was like Huxley in his aggressiveness. But he was classier, so to speak. *Baron* Cuvier was a fashion plate and a speaker who could switch from soft-voiced, lacerating wit to overpowering thunder in a blink. He found the transformist concept of gradual Evolution ludicrous. The much simpler truth was that species were constantly dying out and new ones were taking their place. French naturalists so feared Cuvier's brilliant fury that the Theory of Evolution—like the name Charles Darwin and the ism-magnifying term "Darwinism"—seldom saw print in France...and seldom does, to this day.

In Germany, on the other hand, *The Origin of Species* was an immediate sensation. By 1874 Nietzsche had paid Darwin and his theory the highest praise with the most famous declaration in modern philosophy: "God is dead." Without mentioning Darwin by name, he said the "doctrine that there is no cardinal distinction between man and animal" will demoralize humanity throughout the West; it will lead to the rise of "barbaric nationalistic brotherhoods"—he all but called them by name: Nazism, Communism, and Fascism—and result within one generation in "wars such as never have been fought before." If we take one generation to be thirty years, that would have meant by 1904. In fact, the First World War broke out in 1914. This latter-day barbarism, he went on to say, will in the twenty-first century lead to something worse than the great wars: the total eclipse of all values.[12]

It was not so much the book *The Origin of Species* as the talk about it that created such a stir. The book never came close to becoming a bestseller. Darwin was an engaging writer, but the science was too hard to comprehend. The first printing was only 1,250 copies.* Robert Chambers, still anonymous, reissued *Vestiges* to take advantage of the excitement...and outsold *The Origin of Species* by four or five to one. But Darwin's theory came up with dizzying frequency in newspaper and magazine articles and cartoons— the cartoonists delighted in depicting Darwin with an ape's body—in public debates high (Huxley versus Wilberforce) and low, in doggerel, and, of course, in sermons. No new idea had ever generated so much controversy, gossip, and befuddlement or so many heavy-laden books. By 1863 Darwin's own collection of clippings contained 347 reviews and 1,571 commentaries plus 336 pieces that were never sorted out by category.[13]

It turned out that Leifchild, the unknown *Athenaeum* reviewer who had jumped the gun, was merely the first to assume that Darwin's real subject was the evolution of man—as, of course, it was. Even Darwin's closest allies, Huxley and Lyell, stopped pretending otherwise. In Febru-

---

* Darwin kept detailed notes on the schedule and printing record of *The Origin of Species* in his diary.

ary of 1863, Lyell, who had for so long doubted Evolution, cast his lot with Darwin in a book called *The Antiquity of Man*. The full title was *The Geological Evidences of the Antiquity of Man with Remarks on Theories of the Origin of Species by Variation*. It obliterated the religious distinction between man (made in God's own image) and animal. A few weeks later Huxley published *Evidence as to Man's Place in Nature*. Here was Huxley at his snarling, pugnacious best. He came right out and asserted, in different words, precisely what Nietzsche would be saying eleven years later: there is no cardinal distinction between man and animal. Man descended from animals, and only fools and too-far-gone clergymen could deny it.

It gradually dawned on Darwin that the attacks by Leifchild in the *Athenaeum* and Owen in the *Edinburgh Review* had been godsends. As Sigmund Freud would put it thirty-five years later in similar circumstances, "Many enemies, much honor." Darwin's critics had turned him into a controversial figure, and a very famous one. For years his friends had been fond of him in a good-old-Charlie fashion. But their demeanor and the very expressions on their faces had changed. Suddenly good old Charlie had become a celebrity. No matter what side of the controversy people came down on, no matter how well they knew him, their involuntary smiles in his presence radiated a certain...mousy awe. And oh, yes, Celebrated Old Charlie picked that up every time. Not even his longtime friendly mentor, elder, protector, and superior when it came to social and intellectual status and public recognition, namely, Lyell—not even Lyell could hold back a certain deference. Without a word, both were aware that their rankings had reversed on every score. Darwin was famous. Life was delightful—

—until his next jolt.

Max Müller had been born and educated (at Leipzig University) in Germany but for years had been a professor of modern languages at Oxford and by now, 1861, was the best-known and most distinguished linguist in Britain. He gave two highly publicized lectures at the Royal Institution that year in which he said, apropos of Darwin's Theory of Evolution, "Language is our Rubicon, and no brute will dare to cross it."[14] During the dustup that ensued, he added: "The Science of Language will yet enable us to withstand the extreme theories of evolutionists and to draw a hard and fast line between man and brute."[15]

In *The Origin of Species,* Darwin had dealt only with the Evolution of animals. But his real dream was of being the genius who showed the world that man was just an animal himself, evolved from other animals—and that his mighty mental abilities had evolved the same way. And now, just two years after its publication, an already *certified* genius, Max Müller, was saying man possessed a supreme power—language—that no animal had ever possessed or ever would. They might as well have come from different planets—man, with his power of speech...and animals, with nothing even remotely comparable.

And the accurséd Max Müller just wouldn't let up. He seldom mentioned Darwin by name, but the *world* knew immediately who the target of his mockery was.* Darwin's notion that language had somehow evolved from imitation of animal sounds...Müller called that the *bow-wow* theory. The notion that speech began with instinctive cries such as "Ouch!" for pain and "Oh!" for surprise...Müller called that the *pooh-pooh* theory. The notion that words... "whisper," "wind," "crack," "hack," "belch," and "squash"...came from the sounds things made...Müller called that the *ding-dong* theory. Many Darwinists, such as the highly regarded Sir Richard Paget and George Romanes, didn't seem to realize that Müller was merely making fun of their Prophet with all this baby-talk terminology. So they pitched in and amplified Müller's laugh in their own long faces. Soon there was the *mama* theory, referring to the coos and other nonverbal cues mothers use with infants, later known as "motherese"...the *tata* theory, later known as the "ges-

---

* As early as 1838 Darwin had written in his notebooks and letters about the conundrum of language. Darwin's "Old and Useless Notes About the Moral Sense and Some Metaphysical Points" was transcribed and edited by Paul Barrett and published alongside a facsimile of Darwin's original manuscript in *Darwin Online*. After the publication of *The Origin of Species,* other scholars applied his Theory of Evolution to all aspects of human existence.

tural theory," the notion that humans first communicated via hand signals and body language and—somehow—began to substitute the voice for all the motions...plus the *yo-he-ho* theory...the *sing-song* theory...the *hey, you* theory...there was no end to it...and no end to Max Müller's delight in the Darwinists' wooden noggins.

Soon the biggest butt of Müller's jokes, Darwin himself, with his family and household staff in tow, retreated to the spa, this time in Malvern, for a protracted stay. He was fifty-four. When he emerged two years later, he was an old man, the old man most commonly portrayed in photographs. His dome had gone bald...his hair had turned gray...and he had cultivated a so-called philosopher's beard of the sort that had been the philosopher's status symbol since the days of Roman glory. Darwin was forever pictured sitting slightly slumped in an easy chair...his philosopher's beard lying on his chest all the way from his jaws to his sternum...like a big old hairy gray bib.

In the meantime, Wallace knew nothing about this entire set-to. He had remained in the Malay Archipelago, flycatching for all he was worth. He didn't return to England until 1862, when Max Müller's lectures were published. From the very first, he bowed to make way for Darwin. He referred to *It* as Darwin's Theory of Evolution. He even went so far as to say it was a good thing the theory hadn't been attributed to somebody like himself, Alfred Wallace. It might never have been noticed, whereas it couldn't be ignored once it carried the imprimatur of a Gentleman like Mr. Darwin.* That may have sounded like a queasy form of groveling, but in fact he was absolutely correct.

After his return, the Gentlemen inflated Wallace's reputation for their own purposes—especially Darwin, to assuage his feeling of guilt. Wallace never felt comfortable with any of them except Lyell, who was the old man of the naturalists and had first noticed his talent back in 1855. The others, including Lyell's wife, Mary Horner Lyell, intimidated him. She found Wallace's table manners common, bordering on crude. She and the rest of the Linnean crowd struck him as terribly class-conscious. They were the real thing. They had upper-class drawls—*far* became *fahh, extraordinary* (six syllables and thirteen letters) became *ex-strawwwd-nry* (thirteen letters and three syllables). They excelled at the sort of ironically clever conversation they had picked up and polished at Oxbridge. Even their blandest comments piped out UPPER CLASS! UPPER CLASS! without bringing up the subject of class at all. You could have a conversation with these people and not realize until two weeks later that they had slipped the finest ivory needles between your ribs and insulted you to the core.

Wallace was rendered quiet to the point of vanishing

---

* Wallace says as much in a letter to Darwin dated May 29, 1864.

when he was around them…whereas previously, among others, such as the eccentric Englishman James Brooke, who was the rajah—the actual rajah—of Sarawak, in Borneo, no one had ever been more confident or adept at entertaining a whole roomful of people than Wallace. "The Rajah was pleased to have so clever a man with him," said the rajah's secretary, "and if he could not convince us that our ugly neighbors, the orangutans, were our ancestors, he pleased, delighted and instructed us by his clever and inexhaustible flows of talk—really good talk."[16]

Darwin displayed many symptoms of guilt over nipping Wallace's underwear the way he had. Whenever the discussion, in print or in person, got around to "Mr. Darwin's Theory of Evolution," Darwin always made a point of mentioning that Mr. Wallace had also done important work in this area, so important, in fact, that in 1858 their original papers on the subject had been presented jointly before the Linnean Society. All these references came across as the great master pat pat patting his little protégé on his little head head head again.

Yes, over and over, until the day he died, Darwin sent up flares signaling his guilt. There is a difference, however, between guilt and regret. Of regret the man never betrayed one twitch. Either way, Wallace's reflected light grew brighter and brighter. The accumulation of Darwin's praise (only Lyell and Hooker knew it was guilt-driven) lent Wallace a certain heft, nonetheless…and his constant deferring to Darwin as *the* discoverer of natural selection kept him in the good graces of the Gentlemen.

So Wallace became a celebrity lit up by indirect light. By now that didn't seem to trouble him in the slightest...for it was a *lovely* light. From there on out, nothing he wrote could be ignored. And write he did. He had a wonderfully clear, direct style and seemingly endless energy and originality. He went on to turn out an astonishing seven hundred articles and twenty-two books...he popularized the theory of natural selection—in fact, he wrote a book entitled *Darwinism*—but also branched out into anthropology, geography, geology, and public policy...and he never left Britain to go flycatching again. He was on his way to international renown and enough gold medals from learned societies and Queen Victoria to make his white tie and tails *blaze* with a chestful of honors.

By 1870 Wallace's heft had turned into real gravitas— to Darwin's sudden dismay. Darwin had screwed up his courage and begun working on *The Descent of Man,* his sequel to *The Origin of Species,* formally pronouncing man a descendant of the apes and monkeys and a product of natural selection, when Wallace published "The Limits of Natural Selection as Applied to Man," the end piece of a collection of his articles entitled *Contributions to the Theory of Natural Selection.*

On the thirty-ninth page of this forty-page essay, he gets off a line that lives on in the annals of annihilation by anesthesia. There is nothing in the pages you have just read, says Wallace, that "in any degree affects the truth or the generality of Mr. Darwin's great discovery."[17] No, nothing except for the fact that in the preceding thirty-eight pages he has systematically disassembled and demolished what was dearest to Darwin's heart, the central point of his entire theory from the beginning, namely, that human beings are animals themselves, merely the most highly evolved species of animal, to which Wallace replies, in effect: I'm sorry, but there *is* a cardinal distinction between man and animal.

Once he gets started, Wallace wastes no time moving in for the kill. He goes straight after three of Darwinism's central assumptions. One, natural selection can expand a creature's powers only to the point where it has an advantage over the competition in the struggle for existence— and no further. Two, natural selection can't produce any changes that are bad for the creature. And three, natural selection can't produce any "specially developed organ" that is useless to a creature...or of so little use that it is not until thousands and thousands of years down the line that the creature can take advantage of the organ's full power.

The creature is man, and the "specially developed organ" is the brain. Wallace goes to some pains to demonstrate that among mammals the size of the brain has an "intimate connection" to intelligence. For example, "whenever an adult male European has a skull less than nineteen inches in circumference or has less than sixty-five cubic inches of brain, he is invariably idiotic."[18] The Neanderthals and all other prehistoric human beings dating back to the Stone Age had brains bigger than that and nearly as big as modern man's, as do the most untutored peoples of the present...while the brains of the most intelligent apes, such as chimpanzees, gorillas, and orangutans, are scarcely one-third the size of man's. That means, says Wallace, that prehistoric man had a "specially developed organ" with far more power than he needed to survive...and it was literally ages before modern man began to make full use of it. So here he is, equipped with "an organ that seems prepared in advance, only to be fully utilized as he progresses in civilization."[19]

In no way, said Wallace, can natural selection account for such a thing. But neither can natural selection account for man's hairless body, especially his bare back, which makes him highly vulnerable to wind, cold, and rain. All other primates, even in Africa and the tropics, grow hides or coats of hair that protect them to the point of making them waterproof. The hair of the coats is layered at a downward angle. Rain rolls right off. Does man miss that? All the time, said Wallace. In fact, since time immemorial, men have been using animal hides and anything else they could think of to keep their backs covered.[20] There you had it—an obvious case of what Darwin said couldn't happen: injurious evolution. "A single case of this kind," Darwin himself had said, tempting Fate, "would be fatal to [the] theory."[21]

But there was *fatal*...and there was *smashed* to death, to use Darwin's own word. Smashed to death came in the form of the highest achievement of the human brain: abstract thought. Without that, said Wallace, no man could have conceived of numbers, arithmetic, and geometric forms...he would never have experienced pleasure in music and art...he would have no conscience and therefore no moral codes...he would have no "ideal conceptions of space and time, of eternity and infinity"[22]...no sense of the past or the future...no consciousness of man's place in the world...no capacity for recording the here and now so that he could draw upon accurate memories in making plans for the long term or even five minutes ahead. None of these, mankind's highest and most refined abilities, had anything to do with natural selection. Natural selection could only make a species fit enough to survive, physically, in the struggle for life. *Survival?* Absolute domination is the name for it in man's case. Man's brain "has led to his conquest of the world"...as Wallace put it nineteen years later in his book *Darwinism*. The power of the human brain was so far beyond the boundaries of natural selection that the term became meaningless in explaining the origins of man.

No, said Wallace, "the agency of some other power" was required. He calls it "a superior intelligence," "a controlling intelligence." Only such a power, "a new power of definite character," can account for "ever-advancing" man.[23] Whatever that power is, it is infinitely more important than mere natural selection.

Now, that *hurt*. Once again, this little flycatcher Wallace had (to use an anachronism, as noted above) *freaked out* Charles Darwin. In a regular frenzy Charlie began scrawling *NO!— NO!— NO!— NO!* in the margins of his copy and then hurling spears in the form of exclamation points.[24] Only a few wound up immediately following the *NO*s. The rest of them hit the page in the form of…take *that,* Wallace!…right through your temporal fossa and your little fifty-cubic-inch brain cavity!…and *this* one!— *riiiippp*—right through your solar plexus!…and *this* one!…right through your bowels!…and *this* one!…a regular crotch crusher!…and *this* one…straight through your ungrateful heart!!!!!! And to think that I went to the trouble of building up your reputation. True, it was out of guilt, but I built it up for you all the same. And don't think that pathetic little disclaimer on page 39 absolves you of any treachery, either.

Finally he pulled himself together and sent Wallace a note saying, "I hope you have not murdered too completely your own and my child."[25]

Oh, but he had. Murdered or, less dramatically, tried to destroy what Wallace had done with the entire Olympian climax of Darwin's Theory of Evolution. Darwin was in no mood to be less dramatic, however. Wallace's treachery, coming on top of Max Müller's summary dismissal, *did* qualify as murder, when you got right down to it.

His only relief came when Wallace, bafflingly, began self-destructing as a scientist by becoming a believer in spiritism, which had become something of a vogue among many otherwise intelligent people. Spiritism did not necessarily involve a belief in God. But you did have to assume there was some kind of fourth dimension, the unearthly domain of a force, a spirit, that ordinary mortals couldn't comprehend. This was what Wallace, a confirmed atheist since his early teens, had in mind when he began to go on about "the agency of some other power" ... "a new power of a definite character" ... "a superior intelligence" ... "a controlling intelligence." One way to commune with the Power was to engage in séances, complete with table rapping, tarot cards, and inexplicable moans and cries. One goal, among several, was to get in touch with dead souls on the Other Side of the river. Wallace managed to get Darwin to attend one. He lasted less than fifteen minutes before walking out, shaking his head.[26]

In fact, Wallace was attributing to supernatural powers something as natural as breathing to human beings everywhere—and *only* to human beings—namely, speech, language, the Word.

Language in all its forms advanced man far beyond the boundaries of natural selection, allowing him to think abstractly and plan ahead (no animal was capable of it); measure things and record measurements for later (no animal was capable of it); comprehend space and time, God, freedom, and immortality; and remove items from Nature to create artifacts, whether axes or algebra. No animal could even begin to do any such thing. Darwin's doctrine of natural selection couldn't deal with artifacts, which were by definition unnatural, or with the mother of all artifacts, which was the Word. The inexplicable power of the Word—speech, language—was driving him crazy and Wallace across to the Other Side.

But a cosmogonist like Darwin couldn't let it go at that. Speech had to have some animal genealogy... *had to fit* into his Theory of Everything. It was on his mind constantly. It was a threat he couldn't dodge much longer.

CHAPTER III
THE DARK AGES

IT WAS BAD enough that Max Müller had challenged him directly, flat-out, out loud, in public, before the Royal Institution—and made fun of him—over this accursèd business of language.[1] That was eight years ago, 1861, and all this time Darwin had been trying to find something, *anything,* in the sounds, gestures, habits, and facial expressions of animals that he could claim as evidence of language evolving. It *had to be* there somewhere! But if so, not he, not Lyell, Hooker, Huxley, nor any other of the faithful could find it. And then comes the turncoat Wallace piling on with a whole slew of other powers that seemed to have come to *Homo sapiens* from out of the blue, leaving Darwin and his entire theory and all his devotees stuck on the wrong side of the Rubicon.

At first he had been angry. But now he had a headache.

What a mess. How could he explain his way out of it? He could give in and concede that, well, yes, he hadn't been right about *every*thing, such as this business of language...that way out lasted about one blink. He was too far gone in his cosmogonist obsession. After all, his Theory of Evolution was a theory of...*every*thing.

In a desultory fashion he had begun writing a sequel to *The Origin of Species*. He thought of it as his "Man Book" because it would bring the formerly missing *Homo sapiens* into the big picture of Evolution.* Wallace had already beaten him to the punch on natural selection, and it had taken every bit of Darwin's influence and slick-slick-slickness as a Gentleman to euchre the naive flycatcher out of his priority. But suppose the flycatcher were to come out with a book murdering "your own and my child" and showing that Evolution couldn't possibly account for the tyke's gift *of abstract thought?* What if he put across the idea that some "new power of a definite character," some "controlling intelligence"—some force ordinary mortals couldn't comprehend—set man apart from animals?[2]

As far as Darwin was concerned, the slate on Wallace's roof had come loose...thanks to all this spiritism rubbish. He didn't try to discredit Wallace on those grounds, however. Up to now Darwin's and the Gentlemen's class superi-

---

* Darwin uses this phrase in a June 29, 1870, letter to John Jenner Weir (http://www.darwinproject.ac.uk/entry-7253).

ority had intimidated the flycatcher. But if Wallace got good and riled up, he just might open up the matter of priority again and expose the fast one the trio of Gentlemen—Lyell, Hooker, and Darwin—had put over on him when he was 7,200 miles away in Malaysia and oblivious of what they were up to…besides, any disparagement of this "controlling intelligence" Wallace kept talking about might be interpreted as an attack on the Almighty, and Darwin was already in enough trouble on that score. He had to come up with a more subtle rebuttal to Müller and Wallace. And speed was of the essence. Who knew how far or how fast Wallace might try to go with this stuff?

By then, 1869, Darwin was sixty years old and more of a hopeless dyspeptic, or hypochondriac, than ever. Vomiting three or four times a day had become the usual. His eyes watered and dripped on his old gray philosopher's beard. The chances of his leaving his desk in Down House and going out into the world looking for evidence, as he had on the *Beagle,* were zero. Instead he chained himself to his desk and forced himself to write, as he had in 1858 and 1859, when he forestalled Wallace with *The Origin of Species.* Now he faced the worst threat ever to his Theory of Everything. So he wound his imagination up to the maximum and herded all the animals together in his head, like some Noah the Naturalist, and inspected them—*hyper*inspected them this time—until he found what he was looking for, namely, embryos of all the Higher

Things...language, the moral sense, abstract thought, art, music, religion, self-consciousness...whatever the human mind was capable of, he found early origins of it in animals.[3] The upshot was a real tour de force of literary imagination entitled *The Descent of Man, and Selection in Relation to Sex,* published in 1871.

Thirty-one years later, in 1902, another British writer published another tour de force of literary imagination concerning the origins of animals and man. The writer was Rudyard Kipling, and the book was called *Just So Stories.* A typical story was "How the Leopard Got His Spots." It seems Leopard lived in the barren, dirt-tan, sandy-colored upper reaches of a mountain overlooking the jungle. Leopard's hide was the same color as the terrain, a sandy tan with no markings. Smaller animals didn't even see him until he leaped out of the background and had them for lunch. Leopard's hunting pal was an Ethiopian, a man with light yellowish-brown skin. He used a bow and arrow to turn passersby into mouthfuls...until bad luck drove him and Leopard down into the darkness of the jungle below. Down there sandy-tan Leopard suddenly stood out like a nice bright mouthwatering meal himself...for any pair of incisors that happened by. The Ethiopian wasn't too happy about hanging around with him anymore. To save his own too-light hide, the Ethiopian found some blacking and turned himself black from head to toe. That way he could disappear into the shadows. He had a lot of the black gunk left on his fingers, and so he had a go at Leopard's hide, too, leaving fingerprints all over it. With all the black fingerprints, Leopard looked like nothing more than a pile of rocks on the ground in the jungle's dark green gloom. And that was how the leopard got his spots.[4]

Kipling's intention from the outset was to entertain children. Darwin's intention, on the other hand, was dead serious and absolutely sincere in the name of science and his cosmogony. Neither had any evidence to back up his tale. Kipling, of course, never pretended to. But Darwin did. The first person to refer to Darwin's tales as Just So Stories was a Harvard paleontologist and evolutionist, Stephen Jay Gould, in 1978.[5] Orthodox neo-Darwinists never forgave him. Gould was not a heretic and not even an apostate. He was a simple profane sinner. He had called attention to the fact that Darwin's Just So Stories required a feat of fiction writing Kipling couldn't compete with. Darwin's storytelling power *soared* in *The Descent of Man* precisely where it had to, i.e., in accounting for this perplexing business of language.

Language, said Darwin, originated with the songs birds sang during the mating season. Man began imitating the birds, a cappella. By and by he started repeating certain birdsong sounds so often they began to stand for certain things in nature. They became words in embryo, and man began creating a "musical protolanguage."[6] But mating songs are sung by male birds only. What about human females? No problem. The females started mimicking the males, although in a higher register, and the protolanguage became far more pleasing. In no time the females were talking circles around the males. Female protolanguage, said Darwin, persists to this day…in the form of mothers cooing to their babies. The sounds have no dictionary meaning at all. Yet they signal love, protection, coziness, and mealtime.[7] Anyway, that was "How the Birds Gave Man His Words."

And why is it that *Homo sapiens* was descended from hairy apes but wound up naked—as Wallace had gone to some pains to point out? Even in hottest horrid-torrid Africa, animals such as antelopes had fur to protect them from the wind and rain. So did man…way back in that invisible past, where Evolution lives. Starting out, said Darwin, man was as hairy as the hairiest ape. Why no longer? *Blind,* aren't you, Wallace? You didn't get the second half of my title, *The Descent of Man, and Selection in Relation to Sex,* did you. Evolution, said Darwin, had turned *Homo sapiens* into a more sensitive animal, which in turn gave him something approaching aesthetic feelings. The male began to admire females who had the least apelike hides because he could see more of their lovely soft skin, which excited him sexually. The more skin he saw, the more he wanted to see. Obviously valued by the males because their hides were much less hairy, the most sought-after females began to look down their noses at the old-fashioned hairy males, one crude step away from the apes themselves. Generation after generation went by, thousands of them, until, thanks to natural selection, males and females became as naked as they are today, with but two clumps of hair, one on the head and the other in the pubic area, plus wispy, scarcely visible little remnants of their formerly hirsute selves on the forearms and lower legs and, in the case of some males, the chest and shoulders.* Yes, their backs got cold, terribly cold, as Wallace had argued. But what poor Wallace didn't know was that the heat of passion conquered all...and that was "How Man Lost His Hair Over Love." (Got *that,* Wallace?)

The truth was, Kipling didn't rate an "ism" at the end of his name. Darwin did. When it came to making up stories, Kipling lacked Darwin's great resource, "my dog." For example, how did man obtain the power of abstract thought? Obvious, all too obvious. How can anybody dispute the fact, said Darwin, that even small mammals far below the status of ape have it? "When a dog sees another dog at a distance, it is often clear that he perceives that it is a dog in the abstract, for when he gets nearer his whole manner changes, if the other dog be a friend."[8] Was that his, Darwin's, dog? He doesn't say, but often in *The Descent of Man,* "my dog" steps forth as major evidence. "When I say to my terrier, in

---

* "I am inclined to believe, as we shall see under sexual selection, that man, or rather primarily woman, became divested of hair for ornamental purposes..." Darwin (1871), 149–150.

an eager voice…and I have made the trial many times"—
the "trial" suggesting a controlled scientific experiment—
" 'Hi, hi, where is it?' she at once takes it as a sign that something is to be hunted, and generally first looks quickly all
around, and then rushes into the nearest thicket, to scent
for any game, but finding nothing, she looks up any neigh-
boring tree for a squirrel. Now do not these actions clearly
shew that she had in her mind a general idea or concept that
some animal is to be discovered and hunted?"[9]

Religion? You have but to observe *my dog*. "The feeling
of religious devotion is a highly complex one, consisting
of love, complete submission to an exalted and mysterious
superior, gratitude, hope for the future…" We see "this
state of mind in the deep love of a dog for his master,
associated with complete submission, some fear and perhaps other feelings."[10] He once noticed *my dog* lying on
the lawn on a hot, still day. Not far away "a slight breeze
occasionally moved an open parasol," and *my dog* growled
fiercely and started barking every time. "He must, I think,
have reasoned to himself…that movement without any ap-
parent cause indicated the presence of some strange living
agent…The belief in spiritual agencies would easily pass
into the belief in the existence of one or more gods…A dog
looks upon its master as on a god."[11] And there you have it.
This reverence moves up the great chain of Evolution until
it reaches man.

Parental affection? That begins very low in the animal

hierarchy, with starfish, spiders, and *Forficula auricularia*. *Forficula auricularia* are earwigs in biology-lingo Latin. The moral sense? Parental affection, including the earwigs', is the moral sense in embryo, says Darwin.[12] It has evolved into the sympathy that mammals feel not only for their own kind but also for creatures from other species entirely— even to the point of risking their lives for them. Sympathy "leads a courageous dog to fly at any one who strikes his master...I have myself seen a dog"—*my dog?*—"who never passed a cat who lay sick in a basket without giving her a few licks with his tongue, the surest sign of kind feeling in a dog." He goes on to say, "I saw a person pretending to beat a lady, who had a little timid dog on her lap, and the trial had never been made before"—another scientific trial—and the little creature instantly "jumped away, but after the pretended beating was over it was really pathetic to see how perseveringly he tried to lick his mistress's face and comfort her."[13] Besides love and sympathy, animals exhibit other qualities connected with social instincts, which in man would be called moral. "Dogs," he says, "possess something very like a conscience." Dogs seem to be able to restrain themselves in deference to their master's rules, and "this does not appear to be wholly the result of fear." For example, they will "refrain from stealing food in the absence of their master."[14]

Doggedly, doggedly, Darwin hauls down all Wallace's signs of "a new power of a definite sort" and returns them to the barking, whining, itching, scratching animal life of his Theory of Evolution.

But his great overarching goal was to drain Max Müller's damnable Rubicon dry. If Müller was right or even seemed to be right, that was the end of Darwin's being known as the genius who showed the world that there is no cardinal distinction between man and animal. Language was the crux of it all. If language sealed off man from animal, then the Theory of Evolution applied only to animal studies and reached no higher than the hairy apes. Müller was eminent and arrogant—and made fun of him![15]

The Descent of Man, and Selection in Relation to Sex was not nearly the sensation that Darwin alternately hoped and feared it would be.* His naturalist colleagues, notably Lyell and Huxley, and by then much of the reading public, took it for granted that the real subject of *The Origin of Species* twelve years earlier had been the descent of man from out of the trees, where the monkeys lived. This new book was just filling in the details. By then the Theory of Evolution had won the intellectual status battle, even within the ranks of the Anglican Church's young clergymen. They were turning from clergy into the clerisy themselves. The reviews approached Darwin as an already Great Man. The *Annual*

---

* One anonymous reviewer put it this way: "Mr. Darwin's long expected and lately published volumes will not be so much to startle...as to consolidate, to fortify, and to push to a conclusion the scheme of ideas which the world has learned for years to associate with his name." (See "Darwin's Descent of Man," *Saturday Review* 3, April 1871.)

Register, a yearly survey of British intellectual life, compared him to Isaac Newton, discoverer of the law of gravity and creator of the fields of physics, mechanics, modern astronomy, and the Rules of Scientific Reasoning in the 1600s. *The Register*'s anonymous reviewer said everyone knew "how profound was the influence of the Newtonian philosophy over the next two or three generations." Darwin's theory will have a comparable impact, he predicts. "One comes across traces of its influence in the most remote and unexpected quarters, in historical, social, and even artistic questions...We are everywhere meeting with that series of ideas to which Mr. Darwin has done more than any other man to give prominence."[16]

Darwin's goal was to show that all Müller's and Wallace's Higher Things evolved from animals—animals even as small as earwigs. He had no evidence, causing him to fall back over and over on the life and times of *my dog.* Fellow naturalists, as well as the linguists, seemed less than riveted. This new theory of language prompted no *Ahahh!* responses, let alone *Ahura!* Negative reviews criticized the thinness of his reasoning as well as the lack of evidence, and positive reviews avoided bringing up the subject at all.[17] Obviously Darwin was as baffled about the origin of language as everybody else.

The very next year, 1872, the Philological Society of London gave up on trying to find out the origin of language and would no longer accept papers on the subject or countenance bringing it up at society meetings.[18] The members were getting almost as batty about language as Darwin and Wallace. All the endless cerebration had proved to be pointless. It clarified nothing and drove the society in no direction but into the slough of despond. The Linguistic Society of Paris had banned the subject on the same grounds six years earlier, in 1866.[19]

Of course, philologists and Darwinists were different creatures, as demonstrated when Max Müller took on Darwin in 1861, pooh-poohed him, and declared that language was a sheerly dividing line elevating man above animal in a final and fundamental way. But when Darwin's own attempt, in *The Descent of Man,* failed to clear up the muddle, Darwinists threw their hands up, too. The subject of the origin of language and how it works entered a dark age that was to last for seventy-seven years, comparable in the annals of science to the Dark Ages that descended upon Europe after the invasion of the Huns. In retrospect, it is hard to believe that the most crucial single matter, by far, in the entire debate over the Evolution of man—language—was abandoned, thrown down the memory hole, from 1872 to 1949.

By the time Darwin died, in 1882 at Down House, of a heart attack after almost three months of intermittent chest pains, his great PR army, the X Club, was in a bad way, too, thanks to the same malady, old age. In 1883 one member died of typhoid, and of the remaining eight, only two were healthy enough to continue meeting regularly. One of

the ailing was Huxley, the onetime boy wonder, who suffered from severe recurrent depression and dropped out
for good in 1885, at age sixty (and died ten years later).
A bid to recruit new members was rejected. The X Club,
the most powerful backers any new scientific theory ever
had...passed away unceremoniously in 1893.[20]

More bad news for the Theory of Evolution broke suddenly in 1900, when three different naturalists from three
different countries—Austria, Germany, and the Netherlands—each out to solve, on his own, the mysteries of
biological inheritance, came upon the never-heralded and
long-since-forgotten work of a long-since-deceased contemporary of Darwin, an Austrian monk named Gregor Johann
Mendel. Mendel had been born plain Johann Mendel in
1822 (three years before Huxley) to a pair of landowning
Moravian peasants who realized early on that they had a
prodigy on their hands. They sacrificed themselves down to
the bone for going on fifteen years to pay for his education,
from first grade through a two-year university program he
completed with honors. For whatever reasons, Mendel, like
Huxley, began to suffer bouts of depression and entered an
Augustinian monastery in northern Austria. As was the cus-
tom, he took the vows and was given a new, saintly name,
Brother Gregor, and a typical monastery chore, gardening,
to help provide food for all the brothers.

The gardener had no training in biology, much less
agronomy, but he began to notice certain patterns repeat-

ing in successive generations of pea plants, and in plant life the generations go by rapidly. In 1856 he began an experiment with green peas that took nine years and by and by involved twenty-eight thousand plants, very likely the biggest and longest agricultural experiment up to that time. In 1865 he laid out all the fundamental laws of heredity in a lecture and then in a monograph entitled *Experiments on Plant Hybridization*—and created the modern science of genetics. This was just five years after Darwin's *The Origin of Species.*

Experiments on Plant Hybridization barely made it into a dim German-language journal and wasn't noticed at all anywhere else.* Fortunately for his equanimity, Mendel was an ace self-regarder. He was convinced that his laws of heredity for green peas applied to every living organism, animal as well as vegetable. Darwin died in 1882, unaware of Mendel. Mendel died two years later, in 1884, all but unread but also undaunted. Not long before he died, he wrote himself a note: "I am convinced it will not be long before the whole world acknowledges the results of my work."[21]

Dead he was, and dead right. Sixteen years after he died, an Austrian, a German, and a Dutchman discovered his work in the German journal and became Mendel's posthumous Huxleys.

---

* Mendel sent Darwin a copy of his article, which was found unread with its pages still uncut after both men had died.

Mendelian genetics overshadowed the Theory of Evolution from the very beginning. This new field had come straight out of purely scientific experiments that agronomists and biologists everywhere were able to replicate. The Theory of Evolution, on the other hand, had come out of the cerebrations of two immobile thinkers, one lying on a sweat-wet makeshift bed in a makeshift hut in Malay... thinking...the other behind a stalwart walnut desk in a stately mansion in the countryside near London... thinking...about things no man had ever seen and couldn't even hope to replicate in much less than a few million years. Next to genetic theory, the Theory of Evolution came off not as a science but as a messy guess—baggy, boggy, soggy, and leaking all over the place. Nevertheless, Darwinists had never given up their cosmogonic determination to make Darwinism explain Everything. In the 1920s and 1930s they hit upon the bright idea of co-opting genetics and treating it as one of the Theory of Evolution's components. A component is part of something bigger—right?

That was how the Darwinists made a comeback after forty years as also-rans. Mendel's theory became just one of their tools. Thus was born what came to be known as the modern synthesis. The leading synthesizer was a geneticist from Ukraine, Theodosius Dobzhansky, who had immigrated to the United States in 1927. In 1937 he published the modern synthesis bible, *Genetics and the Origin of Species*...and in 1973, two years before he died, he published a manifesto with a title Darwinists have been quoting ever since: "Nothing in Biology Makes Sense Except in the Light of Evolution."[22]

And nothing about language made sense to Dobzhansky and his modern synthesizers. They pitied—*pitied*—people who still tried to study its origin. It was about as much use as trying to study the origin of extrasensory perception or mental telepathy or messages from the Other Side. To use a *nom de bouffon* from 1959, before the modern synthesis transmuted into neo-Darwinism, any academic who spent time on the origin of language was written off as...a flake.

It was dumbfounding—utterly *dumbfounding!* Three generations of Darwinists and linguists kept their heads stuck in the sand when it came to the origin of the most important single power man possesses. It took a turn of history on the magnitude of World War II to get their attention.

A prominent linguist named Morris Swadesh was the classic case. Before the war he had been a brilliant but thoroughly traditional linguist. In the 1930s he had trekked tirelessly to remote places nobody but the neighbors, if any, had ever heard of—in Mexico, the United States, and Canada, living off coconuts, fava beans, and beef jerky and, in the chronic absence of plumbing, lowering his pants and squatting down in the tall grass...all the while seeking out tribes and other ethnic enclaves few had ever heard of, either...to study their languages...Tarahumara, Purépecha, Otomi, Menominee, Mahican...close to a hundred of them

before he had finished...and sorting them out into language families such as the Algonquin, the Oneida, the
Tarascan...becoming fluent in more than twenty of them
while he was at it.[23] Then World War II broke out. Swadesh
was thirty, well below the military draft's cutoff age of thirty-
five, and wound up in the army assigned to military intelligence projects involving mainline languages—Russian,
Spanish, Chinese, Burmese, and country cousins such as
Burma's Naga language, for use in interpreting, monitoring,
and possibly espionage. (Swadesh was a quick study. He
soaked up so much Naga in one day touring around with a
local guide that he managed to pull off a ten-minute thank-
you speech in the language that night.[24])

The military was not interested in anecdotal material
that academics such as Swadesh had picked up on their
treks. They wanted data suitably mathed-up, quantified,
hardened (the going metaphor), and standardized in the interest of routinized efficiency. All by itself, throughout the
country, the military generated tremendous momentum for
a trend toward an empirical approach. "Empirical" was a
single adjective encompassing all the foregoing (quantified,
hardened, mathed-up, standardized). Empiricism put great
pressure on academics in the soft sciences, such as sociol-
ogy, anthropology, and linguistics, to harden up until they
had at least a ghost of a chance of resembling physics or
chemistry—or, at the very least, biology.

In 1948 Swadesh created, and not very mellifluously

named, a new field of linguistic study: glottochronology…
and its offshoot, glottogenesis…from the Greek *glotto,*
meaning "tongue" and, by extension, "language." Both
terms bristled with lexicostatistical (Swadesh invented the
term) mathematical symbols—

$$\sum_i \frac{(E_i - O_i)^2}{E_i}$$

—such as the sigma, with its sharp angles and bladelike
lines impaling the unwary human brain and making language research look and sound like radioactive carbon dating, which was the whole idea, literally. Radioactive carbon
dating was used to approximate, within thousands of years,
the age of solid objects, chiefly rocks and bones. Swadesh
liked the scientifical look and sound of that. He wanted
to establish the chronology of languages by the changes in
their grammars, syntaxes, spellings, vocabularies, and rates
of absorption of other languages over time, and he wanted
it as hard as radioactive carbon dating appeared to be.[25]

Linguists intrigued by Swadesh's glottogenesis began to
close in on the subject of exactly how language works.
Notable among them was a Canadian sinologist named Edwin G. Pulleyblank:

"Our capacity, through language, to manipulate the mental world and so deal imaginatively with the world of experience," said Pulleyblank, "has been a major factor, perhaps

the major factor, in giving humans the overwhelming advantage over other species in terms of cultural, as opposed to biological, evolution."[26]

Close he was, but he never quite made it to the heart of the matter, which is not merely what language can do but what it *is*...nor did any other glottogenesist.

Swadesh's glottolingo gradually disappeared from the journals.

Swadesh might have been the first, but the most prominent of the war-boom linguists were affiliated with the Massachusetts Institute of Technology. Radar turned that twelve-syllable mouthful into MIT, initials soon uttered worldwide when the subject of brilliant engineering feats came up. In 1940, at the outset of the war, the government set up the Radiation Laboratory at MIT with the urgent, secret, highest-priority mission of developing radar as a military weapon. For one thing, radar could aim bombs from the belly of a bomber toward the target. The program was so successful that at the height of the war in 1945 the Rad Lab—svelted down from nine syllables to two—had 3,897 employees working day and night in an intended-to-be-temporary three-story building known as Building 20, slapped together out of wood framing with asbestos cladding and crammed into the campus only by relocating the squash courts and a cluster of other soft, as it were, facilities. By then Building 20 had 196,200 square feet of floor space, i.e., three and a half football fields' worth.[27]

At war's end the building's radar heroes moved out, and microwave, nuclear science, and communications engineers moved in. Speech communications, as it was called, had become a major discipline, thanks to the war, and a regular *hard science.* The communications angle opened the doors of Building 20 to a pair of soft sciences, too, namely, modern languages and...linguistics. The linguists were now thrust face-to-face with the engineers...and their glow. The engineers were lit up, the entire breed, with the aura of the victorious radar warriors. In 1949 this curious couple, linguistics and engineering, held a joint conference at MIT.[28] Such excitement!—so much of it that from then on the conferences kept rolling in...waves of them. The linguists were now *eager* to be indistinguishable from the engineers. From the very beginning they decorated their papers with enough esoteric equations, algorithms, and that most scientifical of all visual displays, graphs, to outdo even Morris Swadesh and his glottochronologists. And languages? They mastered language after language and wore them proudly, like pelts on a belt.

CHAPTER IV
NOAM CHARISMA

NOBODY IN ACADEMIA had ever witnessed or even heard of a performance like this before. In just five years, 1953–57, a University of Pennsylvania graduate student—a *student,* in his twenties—had taken over an entire field of study, linguistics, and stood it on its head and *hardened* it from a spongy so-called social science into a real science, a *hard* science, and put his name on it: Noam Chomsky.

Officially, on his transcript, Chomsky was enrolled at Penn, where he had completed his graduate school course work. But at bedtime and in his heart of hearts he was living in Boston as a member of Harvard's Society of Fellows and creating a Harvard-level name for himself while he worked on his doctoral dissertation for Penn.*

This moment, the mid-1950s, was the high tide of the "scientificalization" that had become fashionable just after World War II. Get hard! Whatever you do, make it sound scientific. Get out from under the stigma of studying a "social science"! By now "social" meant soft in the brainpan. Sociologists, for example, were to observe and record hour-by-hour conversations, meetings, correspondence, objective manifestations of status concerns, and make the information really hard by converting it into algorithms full of calculus symbols that gave it the look of mathematical certainty—and they failed totally. Only Chomsky, in linguistics, managed to pull it off and turn all—or almost all—the pillow heads in the field rock-hard. Even before receiving his PhD, he was invited to lecture at the University of Chicago and Yale, where he introduced a radically new theory of language. Language was not something you *learned*. You were born with a built-in "language organ." It is functioning the moment you come into the world, just the way your heart and your kidneys are already pumping and filtering and excreting away.[1]

---

* In *Language and Politics* (1988) Chomsky writes that his original dissertation ("which almost no one looked at at the time") was part of a one-thousand-page manuscript he wrote as a graduate student. According to Robert F. Barsky, Chomsky's dissertation committee passed him after reading only one chapter. See *Noam Chomsky: A Life of Dissent* (Cambridge, MA: MIT Press, 1997), 83.

87

To Chomsky, it didn't matter what a child's first language was. Whatever it was, every child's language organ could use the "deep structure," "universal grammar," and "language acquisition device" he was born with to express what he had to say, no matter whether it came out of his mouth in English or Urdu or Nagamese. That was why—as Chomsky said repeatedly—children started speaking so early in life... and so correctly in terms of grammar. They were born with the language organ in place and the power ON. By the age of two, usually, they could speak in whole sentences and generate completely original ones. The "organ"...the "deep structure"...the "universal grammar"...the "device"—as Chomsky explained it, the system was physical, empirical, organic, biological. The power of the language organ sent the universal grammar coursing through the deep structure's lingual ducts to provide nutrition for the LAD, which everybody in the field now knew referred to the "language acquisition device" Chomsky had discovered.[2]

Two years later, in 1957, when he was twenty-eight, Chomsky pulled all this together in a book with the opaque title *Syntactic Structures*—and was on the way to becoming the biggest name in the 150-year history of linguistics. He drove the discipline indoors and turned it upside down. There were thousands of languages on Earth, which to earthlings sounded like a hopeless Babel of biblical proportions.

That was where Chomsky's soon-to-be-famous Martian linguist came in. A Martian linguist arriving on Earth, he often said...often...often...would immediately realize that all the languages on this planet were the same, with just some minor local accents. And the Martian arrived on Earth during almost every Chomsky talk on language.

Only wearily could Chomsky endure traditional linguists who, like Swadesh, thought fieldwork was essential and wound up in primitive places, emerging from the tall grass zipping their pants up. They were like the ordinary flycatchers in Darwin's day coming back from the middle of nowhere with their sacks full of little facts and buzzing about with their beloved multilanguage *fluency,* Swadesh-style. But what difference did it make, knowing all those native tongues? Chomsky made it clear he was elevating linguistics to the altitude of Plato's—and the Martian's—transcendent eternal universals. They, not sacks of scattered facts, were the ultimate reality, the only true objects of knowledge.[3] Besides, he didn't enjoy the outdoors, where "the field" was, and he knew only one language, English. You couldn't very well count the Yiddish and Hebrew he spoke at home growing up. He was relocating the field to Olympus. Not only that, he was giving linguists permission to stay air-conditioned. They wouldn't have to leave the building at all, ever again...no more trekking off to interview boneheads in stench-humid huts. And here on Olympus, you had plumbing.

Chomsky had a personality and a charisma equal to Georges Cuvier's in France in the early 1800s. Cuvier orchestrated his belligerence from sweet reason to outbursts of perfectly timed and rhetorically elegant fury. In contrast, nothing about Chomsky's charisma was elegant. He spoke in a monotone and never raised his voice, but his eyes lasered any challenger with a look of absolute authority. He wasn't debating him, he was enduring him. Something about Chomsky's unchanging tone and visage turned a challenger's power of reason to jelly.

Charismatic figures in their twenties are not a rare breed. In new religious movements they have tended to be the rule, not the exception: Joseph Smith of the Mormons...Siddhartha Gautama, the Buddha...Scientology's David Miscavige, a "prodigy" and L. Ron Hubbard's handpicked successor...the Báb, forerunner of the Baha'i faith...Barton Stone and Alexander Campbell of the International Churches of Christ...the Jehovah's Witnesses' Charles Taze Russell...and Moishe Rosen of Jews for Jesus. Likewise in warfare: the aforementioned Lamarck, a seventeen-year-old enlisted man taking over an infantry company in the midst of battle...Joan of Arc, a French peasant girl who becomes an army general and the greatest heroine in French history—at the age of nineteen...Napoléon Bonaparte, who by the age of twenty-nine had led victories against French royalist forces as well as the Austrians and the Ottoman Empire...Alexander the Great, who had conquered much of the Hellenistic world before his thirtieth birthday...William Wallace, Guardian of Scotland, who at twenty-seven led the Scots to victory over the British at the Battle of Stirling Bridge.

Charismatic leaders radiate more than simple confidence. They radiate authority. They don't tell jokes or speak ironically, except to rebuke—as in "Kindly spare me your 'originality.'" Irony, like plain humor, invariably turns upon some indulgence of human weakness. Charismatic figures show only strength. They refuse to buckle under in the face of threats, including physical threats. They are usually prophets of some new idea or cause.

Chomsky's idea of the "language organ" created great excitement among young linguists. He made the field seem loftier, more tightly structured, more scientific, more conceptual, more on a Platonic plane, not just a huge heaped-up leaf pile of the data field-workers brought in from places one never necessarily heard of before[4]...linguistics would no longer mean working out in the field among more breeds of na—*er...indigenous* peoples...than one ever dreamed existed. Thanks to Chomsky's success, linguistics rose from being merely a satellite orbiting around language studies and became the main event on the cutting edge....The number of full, formed departments of linguistics soared, as did the numbers of field-workers. Fieldwork was no longer a requirement, however, and more linguists than dared confess it were relieved not to have to go into the not-so-great outdoors the Morris Swadesh way. Now all the new, Higher Things in a linguist's life were to be found indoors, at a desk...looking at learnèd journals filled with cramped type instead of at a bunch of hambone faces in a cloud of gnats.

His radical discovery—the language organ—plus his charisma had already advanced Chomsky to the very front of the pack. But what iced it for him was a book review in the journal *Language*. The book was *Verbal Behavior* by B. F. Skinner, the behaviorist psychologist who had supplanted Freud as the leading figure in the pack.[5] Skinner's radical behaviorism, as he called it, had turned Freudianism inside out. Freud sought to get inside his patient's head, all too much like a voodoo houngan or a Gilgamesh, by hearing him recite his dreams—*his dreams!*—plus whatever unspeakable intimate matters were on his conscious mind and interpreting them by using a few (rather simple, as it turned out) pet formulas...e.g., dreams of flying in airplanes and other experiences involving rapid ascents referred to orgasms. Skinner dismissed all this as sheer "mentalism."[6] He wasn't interested in what a patient said or dreamed but in what he *did*, i.e., his observable actions and his behavior, including his verbal behavior.

Every behaviorist finding began in the laboratory with a rat placed in a small chamber known as a Skinner box, a container about the size of a small carton of paper towels with a bar on one wall. The rat sooner or later learns that if it presses the bar, a food pellet drops onto a little tray.

Eventually comes a time when the rat presses the bar...and no pellet drops to the tray. Gradually the rat discovers it will get a pellet only on every *third* press of the bar or some such change of order. Over time the experimenter can keep changing the order until the rat learns to do extraordinary things...such as press the bar, get no pellet, get no reward until it walks in a counterclockwise circle and comes back and presses the bar again. To his great surprise, wrote Skinner, he found that he could "extend these methods to human behavior," even verbal behavior, "without serious modification."[7]

Really? With that, Chomsky clears his throat in the polite scholarly fashion—then pulls out a boning knife and goes to work. Whatever rats and pigeons do in a box, says Chomsky, can be applied to complex human behavior only in "the most gross and superficial way." He goes on: "The magnitude of the failure of [Skinner's] attempt to account for verbal behavior...[is] an indication of how little is really known about this remarkably complex phenomenon."[8]

All Skinner is doing, as Chomsky sees it, is using the technical vocabulary of laboratory experiments—"controls," "probabilities," "stimulus," "response," "reinforcement"— "to creat[e] the illusion of a rigorous scientific theory"...by taking the very same words out of the rat's box and stretching them far enough to fit real-life human beings at large. He loves to sprinkle the statistical term "probabilities" like salt and pepper on his prose to give it the tang of statistical accuracy...when in fact he has stretched those rigorous terms out so far that "probabilities" means nothing more than "probably" or, even lamer, "possibly." The rigorous statistical term "controls" stretches out to a weak, thin "affects." "Stimulus" winds up as a wan "to begin with," and so on, or else they result in nothing at all and turn out to be totally "empty."[9]

Skinner, says Chomsky, seems to believe that the way to maximize a point in science is to generate, in Skinner's own words, "additional variables to increase its probability" and strength. "If we take this suggestion quite literally," says Chomsky, "the degree of confirmation of a scientific assertion can be measured as a simple function of the loudness, pitch, and frequency with which it is proclaimed, and a general procedure for increasing its degree of confirmation would be, for instance, to train machine guns on large crowds of people who have been instructed to shout it."[10]

"A typical example of 'stimulus control' for Skinner," he says, "would be the response to a piece of music with the utterance 'Mozart' or to a painting with the response 'Dutch.' These responses are asserted to be 'under the control of extremely subtle properties' of the physical object or event. Suppose instead of saying 'Dutch' we had said 'Clashes with the wallpaper,' 'I thought you liked abstract work,' 'Never saw it before,' 'Tilted,' 'Hanging too low,' 'Beautiful,' 'Hideous,' 'Remember our camping trip last summer?,' or whatever else might come into our minds when looking at a picture...Skinner could only say that each of these responses is under the control of some other stimulus property of the physical object...This device is simple as it is empty."[11]

As Chomsky grinds through Skinner for twenty thousand words, he uses the expressions "empty," "quite empty," "quite false," "completely meaningless," "perfectly useless," and the like repeatedly...plus "vacuous"..."complete retreat to mentalistic psychology"... "mere paraphrases for the popular vocabulary" (appears on the same page as "perfectly useless," "vacuous," and "likewise empty")... "serious delusion"..."of no conceivable interest"..."play-acting at science"..."This is simply not true"..."no basis in fact"... "very implausible speculation"... "entirely pointless and empty..."[12]

...*empty empty empty* until there is scarcely a single point Skinner makes in *Verbal Behavior* that Chomsky has not exploded into hot air. With this one review he demolished the book, dug the ground out from under the theory of behaviorism (it never got back on its feet), and consigned B. F. Skinner to history. No one in academia had ever seen such a show of power. Chomsky said many years later that from the very beginning his aim had been to reduce behaviorism to an absurdity. And that he did. Noam Chomsky became a power nobody in the field dared trifle with. In the one recorded instance of someone confronting him over this business of a language organ, Chomsky finessed his way out of it con brio. The writer

John Gliedman asked Chomsky the Question. Was he saying he had found a part of human anatomy that all the anatomists, internists, surgeons, and pathologists in the world had never laid eyes on?

It wasn't a question of laying eyes on it, Chomsky indicated, because the language organ was located inside the brain.[13]

Was he saying that one organ, the language organ, was inside another organ, the brain? But organs are by definition discrete entities. "Is there a special place in the brain and a particular kind of neurological structure that comprises the language organ?" asked Gliedman.

"Little enough is known about cognitive systems and their neurological basis," said Chomsky. "But it does seem that the representation and use of language involve specific neural structures, though their nature is not well understood."[14]

It was just a matter of time, he suggested, before empirical research substantiated his analysis. He appeared to be on the verge of the most important anatomical discovery since William Harvey's discovery of the human circulatory system in 1628.

By 1960 Noam Chomsky's reign in linguistics was so supreme, it reduced other linguists to filling in gaps and supplying footnotes for Noam Chomsky. As for any random figure of note who persisted in challenging his authority, Chomsky would summarily dismiss him as a "fraud," a "liar," or a "charlatan." He called B. F. Skinner,[15] Elie Wiesel,[16] Jacques Derrida,[17] and "the American intellectual community"[18] frauds. He called Alan Dershowitz,[19] Christopher Hitchens,[20] and Werner Cohn[21] liars. He pinned the *charlatan* tag on the famous French psychiatrist Jacques Lacan[22]...and he would pin another later on[23]...

Not really very nice—but at least he woke everybody in the field up from the seventy-seven-year coma. All at once academics, even anthropologists and sociologists, discovered the subject of linguistics. Chomsky had provided them the entire structure, anatomy, and physiology of language as a system.

But there remained this baffling business of figuring out just *what it was*—the creation of *the words* themselves, the specific sounds and how they were fitted together, the mechanics of the greatest single power known to man...*How do people do it?*...and their eyes opened wide as if nobody had ever thought of it before. What would eventually become thousands of articles and conference papers began chundering forth, so many that Müller would have run out of doggy sounds to make fun of them with.

One of the most revealing examples of Chomsky's power was the linguist Alvin Liberman's presentation of his *motor theory,* concerning the visual interpretations that affect face-to-face speech. Liberman didn't buy Chomsky's "language organ" for a moment. It took him several years to work up the nerve to say publicly what he really thought.*

The linguists William Stokoe of Gallaudet University (for the deaf), Gordon Hewes, and Roger Westcott edited one of the classics of late-twentieth-century liguistics, *Language Origins,* in 1974—with the proud claim that they had filled in a gap in Chomsky's *Syntactic Structures.***

And on they came, linguists and anthropologists intent upon shoring up Chomsky's great edifice with evidence... the *gestural theory*...the *big brain theory*...the *social complexity theory*...and...and...

...and more and more scholars sat at their desks just like junior Chomskys trying to solve the mysteries of language with sheer brainpower. The results were not electrifying. Nevertheless, Chomsky had brought the field back to life.

In February of 1967—*bango!*—Chomsky shot up clear through the roof of their little world of linguistics and lit up the sky...with a twelve-thousand-word excoriation of America's role in the war in Vietnam entitled "The Responsibility of Intellectuals." The *New York Review of Books,* the

---

* In "An Oral History of Haskins Laboratories," Liberman's colleagues Frank Cooper and Katherine Harris remember developing the motor theory with Liberman. Later in the same interview, Harris describes a 1964 paper as "the first direct attack on Chomsky and Halle." A transcript of the interview, conducted by Patrick W. Nye in 1988, is available on the Haskins Laboratories website at http://www.haskins.yale.edu/history/OH/HL_Oral_History.pdf.

** By extending generative grammar to sign language, the editors supported some of Chomsky's theories but disagreed with others.

most fashionable organ of the New Left in the Vietnam era, published it as a special supplement.[24]

The piece delivered a shock beyond even Chomsky's never-modest expectations. From the very first paragraph to the last, he tore into the United States' "capitalist" rulers, its supine press, its by turns apathetic and pliable intellectuals. He rolled the country over like a big soggy log, exposing the rot rot rot rot on the underside. He accused the United States of "vicious terror bombings of civilians, perfected as a technique of warfare by the Western democracies and reaching their culmination in Hiroshima and Nagasaki, surely among the most unspeakable crimes in history." And Vietnam? "We can hardly avoid asking ourselves to what extent the American people bear responsibility for the savage American assault on a largely helpless rural population in Vietnam, still another atrocity in what Asians see as the 'Vasco da Gama era'"—meaning imperialist—"of world history. As for those of us who stood by in silence and apathy as this catastrophe slowly took shape over the past dozen years—on what page of history do we find our proper place? Only the most insensible can escape these questions...."

"It is the responsibility of intellectuals," he said, "to speak the truth and to expose lies. This, at least, may seem enough of a truism to pass over without comment. Not so, however. For the modern intellectual, it is not at all obvious."

This was an angry god raining fire and brimstone down not merely upon worldlings committing beastly crimes but also upon the anointed angels who had grown soft, corrupt, and silent to the point of complicity with the very forces of Evil it is their sacred duty to protect mankind from.

It was this rebuke of the intellectuals that turned "The Responsibility of Intellectuals" into more than just a provocative essay by an eminent linguist. It became an *event,* an event on the magnitude of Émile Zola's *J'Accuse* in 1898, during the Dreyfus affair in France...when Georges Clemenceau, a radical socialist (later prime minister of France—twice), turned the adjective "intellectual" into a noun: "the intellectual." At that point "the intellectuals" replaced the old term "the clerisy." Zola, Anatole France, and Octave Mirbeau were *the intellectuals* uppermost in Clemenceau's mind, but he by no means restricted that honorific to writers. Anyone involved in any way in the arts, politics, education—even journalism—who discussed the Higher Things from an at least vaguely savory socialist point of view qualified. So from the very beginning *the intellectual* was a hard-to-define, in fact rather blurry, figure who gave off whiffs—at least that much, whiffs—of Left-aware politics and alienation of some sort.

Chomsky proved to be perfect for the role, and not just because of his academic charisma. More important was timing. He knew how to exploit a tremendous stroke of luck: *another war!*—this one in a little country in Southeast Asia. It was a small war compared to World War II, but the jolt it gave universities and colleges in America was just as severe. The draft had been reinstated. Male students rose up in protest and the girls tagged along with them and faculty members sang along with them through every last bar of their anthem, "I Feel Like I'm Fixin' to Die Rag" (to be replaced two years later with "Give Peace a Chance"). In 1967 tremendous pressure, social pressure, began to build up among *the intellectuals* to prove they were more than spectators in the grandstand cheering the brave members of the Movement on. The time had come to prove you were an "activist," i.e., a *brave* intellectual willing to leave the office, go to the streets, and take part in antiwar demonstrations. The pressure on figures like Chomsky, who was only thirty-eight, was intense. He did his part, left the building, and marched in the most publicized demonstration of all, the March on the Pentagon in 1967. He proved he was the real thing. He got himself arrested and wound up in the same cell with Norman Mailer,[25] who was an "activist" of what was known as the Radical Chic variety. A Radical Chic protester got himself arrested in the late morning or early afternoon, in mild weather. He was booked and released in time to make it to the Electric Circus, that year's New York nightspot of the century, and tell war stories. Chomsky founded an organization called Resist and got himself arrested so many times that his wife was afraid MIT would finally get tired of it and can him. She began studying linguistics herself, in case she had to start teaching in order to keep bodies and souls together in the family.

No one seemed to realize it, but the antiwar movement had brought out in Chomsky some real-enough political convictions from his childhood, ideas long since dried up and irrelevant—one would have thought. Chomsky was born and raised in Philadelphia, but his parents were among tens of thousands of Ashkenazic Jews who fled Russia following the assassination of Czar Alexander II in 1881.[26] Jewish anarchists were singled out (falsely) as the assassins, setting off waves of the bloodiest pogroms in history. On top of that, thousands of Jews were forcibly removed from their homes in Moscow, Saint Petersburg, and adjoining regions and led off, some in chains, to the so-called Pale of Settlement, a geographical ghetto along Russia's western frontier. They risked severe punishment if they ventured beyond the Pale...*pale,* as in the pales of a fence. Even inside the Pale they were restricted from entering cities such as Kiev and Nikolaev, from owning or even leasing property, receiving a college education, or engaging in certain professions. By 1910, 90 percent of Russia's Jews—5.6 million in all—were confined to the Pale.[27]

Anarchism had been a logical enough reaction. The word "anarchy" literally means "without rulers." The Jewish refugees from Russian racial hatred translated that as not merely no more czars...but no more authorities of any sort...no public officials, no police, no army, no courts of law, no judges, no jailers, no banks—*no money*—no financial system at all...in short, no government...and no social classes, either. The dream was of a land made up entirely of communes (not terribly different from the hippie communes of the United States in the 1960s).

A dream it was...a dream...and talk talk talk it was, and endless theory theory theory, until—¡*milagroso!* ¡*maravilla!*—more than half of a major nation, Spain, was taken over by anarchist *cooperativas* during the first two years, 1936–1938, of the Spanish Civil War...when the Loyalists, as they were known, were in power.[28] In 1939 General Francisco Franco and his forces crushed the Loyalists in one of their last strongholds, Barcelona, leading to the memorable gob-of-guilt-in-your-eye cry, "Where were *you* when Barcelona fell?"

Noam Chomsky, all ten years of him, was in Philadelphia when Barcelona fell. He was so worked up about it that it was the topic of his first published article...for the student newspaper of the Deweyite progressive school he went to...a piece in which he denounced Franco as a "fascist."[29] His political outlook—anarchism—appears to have been set, fixed forever, at that moment. Or perhaps the word is pre-fixed...pre-fixed in a shtetl in Russia half a century before he was born. Then, at thirty-eight years old, he laced "The Responsibility of Intellectuals" with so much Marxist lingo that people took him to be part of the radical Left, if not an outright Communist. But he routinely denounced the Soviet Union and Marxism-Leninism as well as capitalism and the United States. He was above their tawdry battles. An angry god was speaking from a higher plane.

Chomsky's audacity and his exotic Old World, Eastern European slant on life were things most intellectuals found charming, since by then, 1967, opposition to the war in Vietnam had become something stronger than a passion... namely, a fashion, a certification that one had risen above the herd. This set off what economists call the multiplier effect. Chomsky's politics enhanced his reputation as a great linguist, and his reputation as a great linguist enhanced his reputation as a political solon, and his reputation as a political solon inflated his reputation from great linguist to all-around genius, and the genius inflated the solon into a veritable Voltaire, and the veritable Voltaire inflated the genius of all geniuses into a philosophical giant...Noam Chomsky.

Even in academia it no longer mattered whether one agreed with Chomsky's scholarly or political opinions or not...for fame enveloped him like a golden armature.

The superlatives came pouring forth from 1967 on. In 1979 a Sunday *New York Times* review of Chomsky's *Language and Responsibility* (Paul Robinson's "The Chomsky Problem") began: "Judged in terms of the power, range, novelty and influence of his thought, Noam Chomsky is arguably the most important intellectual alive today."[30] In 1986, in the Thomson Reuters Arts & Humanities Citation Index, which tracks how often authors are mentioned in other authors' work, Chomsky came in eighth...in very fast company...the first seven were Marx, Lenin, Shakespeare, Aristotle, the Bible, Plato, and Freud.[31] The *Prospect–Foreign Policy* world thinkers poll for 2005 found Chomsky to be the number one intellectual in the world, with twice the polling numbers of the runner-up (Umberto Eco).[32] In the *New Statesman*'s 2006 "Heroes of Our Time" listings—the heroes being mainly fighters for justice and civil rights who had been imprisoned for the Cause, such as Nelson Mandela, the Nobel Peace Prize winner (1993) who had served twenty-seven years of a life sentence for plotting the violent overthrow of the South African government, and another Nobel winner, Aung San Suu Kyi, who was under house arrest in Myanmar at the time—Chomsky came in seventh.[33] His arrests were of the token variety that seldom caused the miscreant to miss dinner out. But his status made up for the never-lost time. A *New Yorker* profile of Chomsky in 2003 entitled "The Devil's Accountant" called him "one of the greatest minds of the twentieth century and one of the most reviled."[34] In 2010 the *Encyclopaedia Britannica* put him on the roster in their book *The 100 Most Influential Philosophers of All Time,* along with Socrates, Plato, Aristotle, Confucius, Epictetus, Saint Thomas Aquinas, Moses Maimonides, David Hume, Schopenhauer, Rousseau, Heidegger, Sartre...in other words, the greatest minds in the history of the world.[35] This wasn't fast company, it was a roster of the immortals.

In his new role as an eminence, Chomsky hurled thunderbolts at malefactors down below, ceaselessly, at an astonishing rate...118 books, with titles such as *Manufacturing Consent: The Political Economy of the Mass Media* (coauthored by Edward S. Herman)...*Hegemony or Survival: America's Quest for Global Dominance...Profit over People: Neoliberalism and Global Order...Failed States* (very much including the United States): *The Abuse of Power and the Assault on Democracy*...an average of 2 books per year...271 articles, at a rate of 4.7 per year[36]...innumerable speaking engagements, which finally got him out of the building and onto airplanes and before podiums far away.

At the same time his output of linguistic papers continued apace, climaxing in 2002 with his and two colleagues' theory of recursion.[37] Recursion consists, he said, of putting one sentence, one thought, inside another in a series that, theoretically, could be endless. For example, a sentence such as "He assumed that now that her bulbs had burned out, he could shine and achieve the celebrity he had always longed for." Tucked inside the one thought beginning "He assumed" are four more thoughts, tucked inside one another: "Her bulbs had burned out," "He could shine," "He could achieve celebrity," and "He had always longed for celebrity." So five thoughts, starting with "He assumed," are folded and subfolded inside twenty-two words...*recursion*...On the face of it, the discovery of recursion was a historic achievement. Every language depended upon recursion—*every* language. Recursion was *the one capability* that distinguished human thought from all other forms of cognition...recursion accounted for man's dominance among all the animals on the globe.

Recursion! ...it was not just a theory, it was a *law!*—just like Newton's law of gravity. Objects didn't fall at one speed in most of the world...but slower in Australia and faster in the Canary Islands. Gravity was a *law* nothing could break. Likewise, recursion! ...it was a newly discovered law of life on earth...*recursion!* ...it was the sort of thing that could lift one up to a plateau on Olympus alongside Newton, Copernicus, Galileo, Darwin, Einstein—Noam Chomsky.

CHAPTER V

WHAT THE FLYCATCHER CAUGHT

BY 2005, NOAM CHOMSKY was flying very high. In fact, very high barely says it. The man was...*in...orbit.* He had made over an entire field of study in his own likeness and put his name on it. If anybody brought up the subject of linguistics, two words inevitably followed: Noam Chomsky. After all, in 2002, so old (at seventy-three) he was already a professor emeritus, he had topped even himself. He had discovered and, as linguistics' reigning authority, *decreed* the Law of Recur—

OOOF!—right into the solar plexus!—a twenty-five-thousand-word article in the August–October 2005 issue of *Current Anthropology* entitled "Cultural Constraints on Grammar and Cognition in Pirahã," by one Daniel L. Everett. Pirahã was apparently a language spoken by several hundred—estimates ranged from 250 to 500—members of a tribe, the Pirahã (pronounced Pee-da*-hannh), isolated deep within Brazil's vast Amazon basin (2,670,000 square miles, about 40 percent of South America's entire landmass). Ordinarily, Chomsky was bored brainless by all those tiny little languages that old-fashioned flycatchers like Everett were still bringing back from out in "the field." But this article was an affront aimed straight at him, by name, harping on two points: first, this particular tiny language, Pirahã, had no recursion, none at all, immediately reducing Chomsky's *law* to just another feature found mainly in Western languages; and second, it was the Pirahã's own distinctive culture, their unique ways of living, that shaped the language—not any "language organ," not any "universal grammar" or "deep structure" or "language acquisition device" that Chomsky said all languages had in common.

It was unbelievable, this attack!—because Chomsky remembered the author, Daniel L. Everett, very well. At least twenty years earlier, in the 1980s, Everett had been a visiting scholar at MIT after working toward an ScD in linguistics from Brazil's University of Campinas (Universidade Estadual de Campinas). He was a starstruck Chomskyite at the

---

* Portuguese speakers pronounce an *r* as a *d* when it begins an interior syllable.

time.* He had an office right across the hall from Chomsky himself. In 1983 Everett received his doctorate from Campinas after writing his dissertation along devout Chomskyan lines, and he didn't stop there. In 1986 he rewrote the dissertation into a 126-page entry in the *Handbook of Amazonian Languages.*[1] It was very nearly an homage to Chomsky. Now that he had his ScD he took periodic breaks in his work with the Pirahã to teach at Campinas, at the University of Pittsburgh as chairman of the linguistics department, and at the University of Manchester in England, where he was professor of phonetics and phonology when he wrote his fateful paper on Pirahã's cultural restraint for *Current Anthropology.*[2]

In his twenty-two years as an off-and-on faculty member, he had written three books and close to seventy articles for learnéd journals, most of them about his work with the Pirahã. But this was his first bombshell. It was one of the ten most cited articles in *Current Anthropology*'s fifty-plus-year history.

The blast set off no *Ahahhs!* let alone *Ahuras!* within the field, however. Quite the opposite. Noam Chomsky and his Chomskyites *were* the field. Everett struck them as a born-again Alfred Russel Wallace, the clueless outsider who

---

* He was. Everett began his academic career in linguistics as a full-fledged Chomsky acolyte. His earliest work aims to apply the Chomskyan model to Pirahã and make excuses for when it didn't quite fit. It took years for him to realize that his adherence to Chomskyan beliefs was preventing him from deciphering Pirahã.

crashes the party of the big thinkers. Look at him! Everett was everything Chomsky wasn't: a rugged outdoorsman, a hard rider with a thatchy reddish beard and a head of thick thatchy reddish hair. He could have passed for a ranch hand or a West Virginia gas driller. But of course! He was an old-fashioned flycatcher inexplicably here in the midst of modern air-conditioned armchair linguists with their radiation-bluish computer-screen pallors and faux-manly open shirts. They never left the computer, much less the building. Not to mention Everett's personal background...he was from a too small, too remote, too hot—it averaged one hundred degrees from June to September and occasionally hit 115— too dusty, too out-of-it California town called Holtville, way down near the Mexican border. His father was a some-time cowboy and all-the-time souse and roustabout. He and Everett's mother had gotten married in their teens and broke up when Everett was not yet two years old. When he was eleven, his mother was in a restaurant staggering beneath a tray full of dirty dishes when she collapsed with a crash and died from an aneurysm.

His father returned from time to time and tried to do his best for his son. His "best" consisted of the lessons of life he taught him, such as taking the boy, who was fourteen at the time, to a Mexican whorehouse to lose his virgin-ity...and then banging on the whore's door and yelling to his son, "Jesus H. Christ, what's keeping you?"...it being *his,* Dad's, turn next.[3]

Helpless, hopeless, the boy went with the flow into the loose louche lysergic life of teenagers in the 1960s. He had just swallowed some LSD in a Methodist church—wondering what it would be like to experience acid zooms amid the curlicued decorations of the sanctuary—when he came upon a beautiful girl named Keren, about his age, with raven hair and ravishing lips. He fell so madly in love—what did it matter that she also had a willpower as blindingly bright and unbending as stainless steel?

She straightened him out very fast. She turned out to be a *real* Methodist. Her mother and father were missionaries. She made a convert out of Everett in no time. Like Everett's own parents, he and Keren got married in their late teens. Keren revved him up to an *evangelical* Methodist, and they resolved to head out into the world as missionaries, like Keren's parents. They underwent several years of intensive linguistic training at the Moody Bible Institute in Chicago, founded by a popular late-nineteenth-century evangelist, Dwight Moody, and the Summer Institute of Linguistics, headed by a later evangelical Christian, Ken Pike. These were tough, rigorous academies, with no fooling around. The Summer Institute's program gave advanced training in various tribal tongues and put the students through four months of survival training for life in the jungle, among other dangerous terrains. The purpose of the Moody Institute and the SIL, as the Summer Institute of Linguistics was called, was to produce missionaries who could convey to prospective converts the Word—the story of Jesus—in their own languages, anywhere on God's earth.*

Everett had turned out to be such a remarkably adept student the SIL encouraged him to see what he could do with the Pirahã, a tribe that lived in isolation way up one of the Amazon's nearly fifteen thousand tributaries, the Maici River. Other missionaries had tried to convert the Pirahã but could never really learn their language, thanks to highly esoteric constructions in grammar, including meaningful glottal stops and shifts in tone, plus a version consisting solely of bird sounds and whistles...to fool their prey while out hunting.[4]

It took three years, but Everett finally mastered it all, even the bird-word warbling, and became, so far as is known, the only outsider who ever did.** Pirahã was a version of the Mura tongue, which seemed to have vanished everywhere else.[5] The Pirahã were isolated geographically. They had no neighbors to threaten them...or change them. It dawned on Everett that he had come upon a people who had preserved a civilization virtually unchanged for thousands, godknew-how-many thousands, of years.

---

* Both the Moody Bible Institute (www.moody.edu) and SIL (www .sil.org) are still in existence.

** In an interview with the *Guardian* Everett explains that it took him one year to get the basics and another two years to be able to communicate effectively. ("Daniel Everett: 'There Is No Such Thing as Universal Grammar,'" by Robert McCrum, March 24, 2012.)

They spoke only in the present tense. They had virtually no conception of "the future" or "the past," not even words for "tomorrow" and "yesterday," just a word for "other day," which could mean either one.[6] You couldn't call them Stone Age or Bronze Age or Iron Age or any of the Hard Ages because the Ages were all named after the tools pre-historic people made. The Pirahã made none. They were pre-toolers. They had no conception of making something today that they could use "other day," meaning tomorrow in this case. As a result, they made no implements of stone or bone or anything else. They made no artifacts at all—with the exception of the bow and arrow and a scraping tool used to make the arrow. So far no one has been able to figure out how the bow and arrow—an artifact if there ever was one—became common to the Inuit (the new "po-litically correct" name for Eskimos) at the North Pole, the Chinese in East Asia, to the Indians—er—Native-born in North America, and the Pirahã in Brazil.

Occasionally some Pirahã would sling together crude baskets of twigs and leaves. But as soon as they delivered the contents, they'd throw the twigs and leaves away.[7] Like-wise...housing. Only a few domiciles had reached the hut level. The rest were lean-tos of branches and leaves. Palm leaves made the best roofing—until the next strong wind blew the whole thing down. The Pirahã laughed and laughed and flung together another one...here in the twen-tieth and twenty-first centuries.[8]

Pirahã was a language with only three vowels (*a, o, i*) and eight consonants (*p, t, b, g, s, h, k,* and *x,* which is the glottal stop). It was the smallest and leanest language known. The Pirahã were illiterate—not only lexically but also visually. Most could not figure out what they were looking at in two-tone, black-and-white photographs, even when they depicted familiar places and faces.[9] In the Pirahã, Everett could see that he had before him the early history of speech and visual deciphering and, miraculously, could study them alive, in the here and now. No such luck with mathematics, however. The Pirahã had none. They had no numbers, not even 1 and 2; only the loose notion of "a little" and "a lot." Money was a mystery to them. They couldn't count and hadn't the vaguest idea of what counting was. Every night for eight months—at *their* request—Everett had tried to teach them numbers and counting. They had a suspicion that the Brazilian river traders, who arrived regularly on the Maici, were cheating them. A few young Pirahã seemed to be catching on. They were beginning to do to real mathematics. The elders sent them away as soon as they noticed. They couldn't stand children making them look bad. So much for math on the Maici. They had to continue paying the traders with vast quantities of Brazil nuts, which they gathered from the ground in the jungle. They were hunter-gatherers, as the phrase goes, but the hunting didn't do them much good in the river trade. They had no clue about smoking or curing meat.[10]

Because they had little conception of "the past," the Pirahã also had little conception of history. Everett ran into this problem when he tried to tell them about Jesus.

"How tall is he?" the Pirahã would ask.

Well, I don't really know, but—

"Does he have hair like you?," meaning red hair.

I don't know what his hair was like, but—

The Pirahã lost interest in Jesus immediately. He was unreal to them. "Why does our friend Dan keep telling us these Crooked-Head stories?" The Pirahã spoke of themselves as the Straight Heads. Everybody else was a Crooked Head, including Everett and Keren—and how could a Crooked Head possibly improve the thinking of a Straight Head? After about a week of Jesus, one of the Pirahã, Kóhoi, said to Everett politely but firmly, "We like you, Dan, but don't tell us any more about this Jesus." Everett paid attention to Kóhoi. Kóhoi had spent hours trying to teach him Pirahã. Neither Everett nor Keren ever converted a single Pirahã. Nobody else ever did, either.[11]

The Pirahã had not only the simplest language on earth but also the simplest culture. They had no leaders, let alone any form of government. They had no social classes. They had no religion. They believed there were bad spirits in the world but had no conception of good ones. They had no rituals or ceremonies at all. They had no music or dance whatsoever. They had no words for colors. To indicate that something was red they would liken it to

blood or some berry. They made no jewelry or other bod-
ily ornaments. They did wear necklaces...lumpy asym-
metric ones intended only to ward off bad spirits. Aes-
thetics played no part—not in dress, such as it was; not
in hairstyles. In fact, the very notion of *style* was foreign
to them.[12]

Here, now, in the flesh, was the type of society that
Chomsky considered ideal, namely, *anarchy,* a society perfectly free from all the ranking systems that stratified and
stultified modern life. Well...here it is! Go take a look! If it
left at some unlikely hour before dawn, you could catch an
American Airlines flight from Logan International Airport,
in Boston, to Brasília and from Brasília, a Cessna floatplane
to the Maici River...you could see your dream, *anarchy,*
walking...in the sunset.

Chomsky wasn't even tempted. For a start, it would
mean leaving the building and going out into the abomin-
able "field." But mainly it would be a triumph for Everett
and a humiliation for himself, headlined:

Everett to Chomsky:
COME MEET THE TRIBE
THAT KO'D YOUR THEORY

Chomsky never willingly mentioned Everett by name after that, nor did he expound upon the Amazon tribesmen
everybody else in linguistics and anthropology was sud-

denly talking about. Chomsky didn't want to know. He didn't particularly want to hear about the Pirahã lore that so fascinated other people, such as the way they said good night, which was "Don't sleep—there are snakes."

And there *were* snakes...anacondas thirty feet long and weighing five hundred pounds, often lurking near the banks in the shallows of the Maici, capable of coiling themselves around jaguars—and humans—and crushing them and swallowing them whole...lancehead pit vipers, whose bite injects a hemotoxin that immediately causes blood cells to disintegrate and burst, making it one of the deadliest snakes in the world...heavy-bodied tree boas that can descend from the branches above and suffocate human beings...plus various deadly amphibians, insects, and bats...black caimans, which are gigantic alligators up to twenty feet long with jaws capable of seizing monkeys, wild pigs, dogs, and now and again humans and forcing them under water to drown them and then, like anacondas, swallowing them whole...Brazilian wandering spiders, as they are called, if not *the* most venomous spiders on earth, close to it...golden poison dart frogs—*poisonous frogs!*—swollen with enough venom to kill ten humans...inch-long cone-nose assassin bugs, also known as kissing bugs because of their habit of biting humans on the face, transmitting Chagas' disease and causing about 12,500 deaths a year...nocturnal vampire bats that can drink human blood for as long as thirty minutes at a time while their human victims sleep.

Walking barefoot or in flip-flops at night in Pirahã land was a form of Russian roulette...and so the Pirahã had learned to be light sleepers. Long middle-of-the-night conversations were not uncommon, so wary were they throughout the midnight hours.

Whatever else it was, Everett's twenty-five-thousand-word revelation of life among the Pirahã was sensational news in 2005. He had decided not to publish it in any of the leading linguistics journals. Their circulations were too small. Instead he chose *Current Anthropology,* which was willing to publish the entire twenty-five thousand words, uncut. That took up a third of the August–October 2005 issue and included eight formal comments solicited from scholars around the world—France, Brazil, Australia, Germany, the Netherlands, the United States.* Two of the scholars, Michael Tomasello and Stephen Levinson, were affiliated with the prestigious Max Planck Institute. By no means were their comments—or any others—valentines. They all had their reservations about this and that. So much the better. The big academic presentation paid off. Radio, television, and the popular press picked up on it here and abroad. Germany's biggest and most influential magazine, *Der Spiegel,* said the Pirahã, a "small hunting and gathering tribe, with a population of only 310 to 350, has become the center of a raging

---

* The complete list of commenters: Brent Berlin, Marco Antonio Gonçalves, Paul Kay, Stephen Levinson, Andrew Pawley, Alexandre Surrallés, Michael Tomasello, and Anna Wierzbicka.

debate between linguists, anthropologists and cognitive researchers. Even Noam Chomsky of the Massachusetts Institute of Technology and Steven Pinker of Harvard University, two of the most influential theorists on the subject, are still arguing over what it means for the study of human language that the Pirahã don't use subordinate clauses."[13]

The British newspaper the *Independent* zeroed in on recursion. "The Pirahã language has none of [recursion's] features; every sentence stands alone and refers to a single event....Professor Everett insists the example of the Pirahã, because of the impact their peculiar culture has had upon their language and way of thinking, strikes a devastating blow to Chomskyan theory. 'Hypotheses such as universal grammar are inadequate to account for the Pirahã facts because they assume that language evolution has ceased to be shaped by the social life of the species.'" The Pirahã's grammar, he argues, comes from their culture, not from any preexisting mental template.[14]

The *New Scientist* said, "Everett also argues that the Pirahã language is the final nail in the coffin for Noam Chomsky's hugely influential theory of universal grammar. Although this has been modified considerably since its origins in the 1960s, most linguists still hold to its central idea, which is that the human mind has evolved an innate capacity for language and that all languages share certain universal forms that are constrained by the way that we think."[15]

In academia scholars are supposed to think and write at a level far above the excitement of the popular media. But Everett and his Pirahã publicity got so deeply under the scholars' skin, they couldn't stand it any longer. In 2006, MIT's linguistics department—not Noam Chomsky's linguistics department—invited Everett to give a lecture about the "cultural factors" that made the Pirahã and their language so exceptional. Three days beforehand, a diatribe appeared on all the Listservs usually reserved for notices about talks to the MIT linguistics community, calling Everett a shameless out-and-out liar who falsifies evidence to support his claims concerning the Pirahã and their language. In fact, says the writer, Everett is so utterly shameless that he had already written about this small Amazonian tribe twenty years earlier in his doctoral dissertation . . . and is now blithely and brazenly contradicting himself whenever he feels like it. I'm publishing all this ahead of time, says the writer, for fear I and others who see through Everett's scam will be "cut off" if we try to expose him at the event itself. In his peroration he says, eyeteeth oozing with irony:

"You, too, can enjoy the spotlight of mass media and closet exoticists! Just find a remote tribe and exploit them for your own fame by making claims nobody will bother to check!"[16] It turned out to be by Andrew Nevins, a young, newly hired linguist at Harvard. He couldn't hold it in any longer!

Nobody in the used-to-be-seemly field of linguistics or any other discipline had ever seen a performance like this before. Even Chomsky's execration of B. F. Skinner had maintained a veneer of politeness and scholarly protocol.

When Tom Bartlett of the *Chronicle of Higher Education* e-requested an interview, Nevins e-replied:

"I may be being glib, but it seems you've already analyzed this kind of case!" Below Nevins's message was a link to an article Bartlett had written about a Dutch psychologist who had confessed to fabricating results by citing studies that had never been made, i.e., were sheer fiction. Bartlett invited Nevins to expand on the implication that Everett was trying to pull off a hoax. Nevins replied, the "world does not need another article about Dan Everett.[17]

What he actually meant, it turned out, was, "The world needs just one more article about Dan Everett, and I'm writing it." Nevins was already at work with two other linguists, David Pesetsky and Cilene Rodrigues, on an article so long—31,000 words—that it was the equivalent of well over 110 pages in a dense, scholarly book.[18] They fought Everett point by point, no matter how dot-size the point. The aim, obviously, was to carpet bomb, obliterate, every syllable Everett had to say about this miserable little tribe he claimed he had found somewhere in the depths of Brazil's Amazon basin. It appeared online as "Pirahã Exceptionality: A Reassessment," by "Andrew Nevins (Harvard), David Pesetsky (MIT), and Cilene Rodrigues (Universidade Estadual de Campinas)"...three linguists from three different universities, Pesetsky pointed out[19]...*hmmm*...a bit... disingenuously...because put them all together...they spelled CHOMSKY (MIT). Chomsky had been David Pesetsky's dissertation supervisor when Pesetsky got his doctoral degree at MIT in 1983.[20] Five years later he returned as Chomsky's junior colleague on the linguistics faculty. Chomsky's close friend Morris Halle, the MIT linguist who back in 1955 had played a major role in bringing him to MIT in the first place, became the dissertation supervisor to Andrew Nevins. Nevins was an MIT lifer. He had enrolled as a freshman in 1996 and had been there for nine years by the time he received his PhD in 2004[21]...and married Cilene Rodrigues, a Brazilian linguist who had been a visiting scholar at MIT for several of the past four years. What they wrote, "Pirahã Exceptionality: A Reassessment," couldn't have seemed more of a Chomsky production had he put his byline on it.

The problem was, it had taken the truth squad, namely, Nevins, Pesetsky, and Rodrigues, all of 2006 to assemble this prodigious weapon. They planned to submit it to the biggest and most influential linguistics journal, *Language,* but it could easily take another six or eight months for *Language* to put it through their meticulous review process. So the trio first decided to publish it online on LingBuzz, a linguistics article-sharing site with a large Chomsky following. Their behemoth doomsday rebuttal appeared there on March 8, 2007—

—and keeled over thirty-three days later, April 10. On that day, the *New Yorker* published a ten-thousand-word piece about Everett entitled "The Interpreter: Has a Remote Amazonian Tribe Upended Our Understanding of Language?" by John Colapinto, with a subhead reading "Dan Everett believes that Pirahã undermines Noam Chomsky's idea of a universal grammar." The magazine had sent the writer, Colapinto, down to the Amazon basin with Everett.

In his opening paragraph Colapinto describes how he and Everett arrived on the Maici in a Cessna floatplane. Up on the riverbank were about thirty Pirahã. They greeted him with what "sounded like a profusion of exotic songbirds, a melodic chattering scarcely discernible, to the uninitiated, as human speech." Colapinto's richest moment came when the linguist W. Tecumseh Fitch arrived. Fitch was a reverent Chomskyite. He had collaborated with Chomsky and Marc Hauser in writing the 2002 article proclaiming Chomsky's discovery that recursion was the very essence of human language. Fitch wanted to see the Pirahã for himself, and Everett had said come right ahead. Fitch had devised a test by which he somehow—it was all highly esoteric and superscientifical—could detect whether a person was using "context-free grammar" by filming his eye movements while a cartoon monkey moved this way and that on a computer screen, accompanied by simple audio cues.

He was absolutely sure the Pirahã would pass the test. "They're going to get this basic pattern. The Pirahã are humans—humans can do this."

Fitch was very open about why he had come all the way from Scotland into the very bowels of the Amazon basin: to prove that, like everybody else, the Pirahã used recursion. At the University of St. Andrews he had left the building a few times to do fieldwork on animal behavior, but never for anything even remotely like this: to study an alien tribe of human beings he had never heard of before...well beyond the boundary line of civilization, law and order, in the rain-forests of Brazil's wild northwest.

With Everett's help he set up a site for his experiments, complete with video and audio equipment. The first subject was a muscular Pirahã with a bowl-shaped haircut. He did nothing but look at the floating monkey head. He ignored the audio cues.

"It didn't look like he was doing premonitory looking," i.e., trying to sense what the monkey might do, Fitch said to Everett. "Maybe ask him to point to where he thinks the monkey is going to go."

"They don't point," Everett said. And they don't have words for left or right or over there or any other direction. You can't tell them to go up or down; you have to say something concrete such as "up the river" or "down the river." So Everett asked the man if the monkey was going upriver or downriver.

The man said, "Monkeys go to the jungle."

Fitch has been described as a tall, patrician man, very much the old Ivy League sort. His full name is William Tecumseh Sherman Fitch III. He is a direct descendant of William Tecumseh Sherman, the famous Civil War general.* But now with Everett in the Amazon basin, he was sweating, and his brow was beginning to fold into rivulets between his eyebrows and on either side of his nose. He ran the test again. After several abortive tries, Fitch's voice took on "a rising note of panic." "If they fail in the recursion one—it's not recursion; I've got to stop saying that. I mean embedding. Because, I mean, if he can't get *this*—"

In the Amazon basin, the tall patrician is reduced to ejaculations such as "Fuck! If I'd had a joystick for him to *hunt* the monkey!" He departs, insisting to Colapinto that his experiments have been a success. But when Colapinto asks him in what ways, his diction turns to fog. Fitch reports to Chomsky in due course that he did detect "context-free grammar" in Pirahã...even though you had to listen and watch the monkey closely, as closely as a Nevins or a Cilene Rodrigues would, to pick it up. As for "context-free" *syntax,* those results were inconclusive.[22]

The *New Yorker* piece made Chomsky furious. It threw

* According to Fitch's curriculum vitae, General Sherman was his great-great-great-grandfather. The general served in the Union army and is best known for his March to the Sea, which cut a sixty-mile-wide strip of wreckage and horror through Georgia to claim Atlanta and Savannah for the North.

him and his followers into full combat mode. He had turned down Colapinto's request for an interview, apparently to position himself as aloof from his challenger. He and Everett were not on the same plane. But now *the whole accursèd world* was reading the *New Yorker*. *Dan* Everett, the *New Yorker* called him, *Dan,* not *Daniel L. Everett*...in the magazine's eyes he was an instant folk hero...Little Dan standing up to daunting Dictator Chomsky.

In the heading of the article was a photograph, reprinted many times since, of Everett submerged up to his neck in the Maici River. Only his smiling face is visible. Right near him but above him is a thirty-five-or-so-year-old Pirahã sitting in a canoe in his gym shorts. It became the image that distinguished Everett from Chomsky. Immersed!—up to his very neck, Everett is...immersed in the lives of a tribe of hitherto unknown na—*er*—indigenous peoples in the Amazon's uncivilized northwest. No linguist could help but contrast that with everybody's mental picture of Chomsky sitting up high, very high, in an armchair in an air-conditioned office at MIT, spic-and-span...he never looks down, only inward. He never leaves the building except to go to the airport to fly to other campuses to receive honorary degrees...more than forty at last count...and remain unmuddied by the Maici or any of the other muck of life down below.

Not that Everett in any way superseded Chomsky. He was far too roundly resented for that. He was telling academics that they had wasted half a century by subscribing to Chomsky's doctrine of Universal Grammar. Languages might appear wildly different from one another on the surface, Chomsky had taught, but down deep all shared the same structure and worked the same way. Abandoning that Chomskyan first principle would not come easily.

That much was perhaps predictable. But by now, the early twenty-first century, the vast majority of people who thought of themselves as intellectuals were atheists. Believers were regarded as something slightly worse than hapless fools. And the lowest breed of believers was the evangelical white Believer. There you had Daniel Everett. True, he had converted from Christianity to anthropology in the early 1980s—but his not merely evangelical but missionary past was a stain that would never fade away completely...not in academia.

Even before the term "political correctness" entered the language, linguists and anthropologists were careful not to characterize any—*er*—indigenous peoples as crude or simpleminded or inferior in any way. Everett was careful and a half. He had come upon the simplest society in the known world. The Pirahã thought only in the present tense. They had a limited language; it had no recursion, which would have enabled it to stretch on endlessly in any direction and into any time frame. They had no artifacts except for those bows and arrows. Everett bent over backwards to keep the Pirahã from sounding the least bit crude or simpleminded. Their language had its limits—but it had a certain profound richness, he said. It was the most difficult language in the world to learn—but such was the price of complexity, he said. Everett expressed nothing but admiration when it came to the Pirahã. But by this time, even giving the vaguest hint that you looked upon some—*er*— indigenous peoples as stone simple was no longer elitist. The word, by 2007, was "racist." And *racist* had become hard tar to remove.

Racist...out of that came the modern equivalent of the Roman Inquisition's declaring Galileo "vehemently suspect of heresy" and placing him under house arrest for the last eight years of his life, making it impossible for him to continue his study of the universe. But the Inquisition was at least wide open about what it was doing. In Everett's case, putting an end to his life's work was a clandestine operation. Not long after Colapinto's *New Yorker* article appeared, Everett was in the United States teaching at Illinois State University when he got a call from a canary with a PhD informing him that a Brazilian government agency known as FUNAI, the Portuguese acronym for the National Indian Foundation, was denying him permission to return to the Pirahã...on the grounds that what he had written about them was...*racist*. He was dumbfounded.

Now he was convinced that the truth squad was waging outright war. He began writing a counterattack faster than he had ever written anything in his life. He didn't know, but wouldn't have been surprised to learn, that Nevins, Peset-sky, and Rodrigues were already at work, converting their online carpet bomb on LingBuzz into a veritable hecatomb to run in *Language* and snuff out Everett's heresy once and for all.

There was no rushing *Language*'s editors, however. They found the piece too long. By the time the squad rewrote the piece...and *Language,* never in a hurry, edited it...and the article, bearing the old LingBuzz title, "Pirahã Exception-ality: A Reassessment," seemed far enough along to make *Language*'s May 2009 issue[23] —

—Everett executed a *coup de scoop.*

CHAPTER VI

THE FIREWALL

IN NOVEMBER OF 2008, a full seven months before the truth squad's scheduled hecatomb time for Everett, he, the scheduled mark, did a stunning thing. He maintained his mad pace and beat them into print—with one of the handful of popular books ever written on linguistics: *Don't Sleep, There Are Snakes,* an account of his and his family's thirty years with the Pirahã.[1] It was dead serious in an academic sense. He loaded it with scholarly linguistic and anthropological reports of his findings in the Amazon. He left academics blinking...and nonacademics with eyes wide open, staring. The book broke free of its scholarly binding right away.

Margaret Mead had her adventures among the Samoans, and Bronislaw Malinowski had his among the Trobriand Islanders. But Everett's adventures among the Pirahã kept blowing up into situations too deadly to be written off as "adventures."

There were more immediate ways to die in the rainforests than anyone who had never lived there could possibly imagine. The constant threat of death gave even Everett's scholarly observations a grisly edge…especially compared to those of linguists who never left their aerated offices in Cambridge, Massachusetts.

In the rainforests, mosquitoes transmitting dengue fever, yellow fever, chikungunya, and malaria rose up by the cloudful from dusk to dawn, as numerous as the oxygen atoms they flew through, or so it felt. No matter what precautions you took, if you lived there for three months or more, you were guaranteed infection by mosquitoes penetrating your skin with their proboscises' forty-seven cutting edges, first injecting their saliva to prevent the puncture from clotting and then drinking your blood at their leisure. The saliva causes the itching that follows.

In 1979, barely a year into the Everetts' thirty years with the Pirahã, Keren and their older daughter, Shannon, came down with high fevers, the shakes, the chills, the itches, the whole checklist from back when Everett once had typhoid fever. So for five days he treated them with antibiotics from his missionary medicine kit, as instructed. The fevers did not abate. Keren's temperature rose to the very tip of the thermometer. Their only hope was to head for the hospital at the provincial capital, Porto Velho, the nearest outpost of civilization, four hundred miles inland on another river, the Madeira.

They set out on the Maici, the entire family—Everett, Keren, Shannon, Kristene, who was four, and Caleb, only two—crammed together in an aluminum canoe Everett had borrowed from a visiting Catholic missionary. All it had was a 6.5-horsepower outboard motor. In a tinny, tiny, tippy canoe overloaded like this, every moment felt like the last moment before capsizing into a jungle river fifty feet deep. Keren was already delirious. She slapped at both Shannon and Everett. It took ten hours to reach the point where they had to cross overland from the Maici to the Madeira. Then, a miracle—the kindness of strangers—four young Brazilians appeared from out of nowhere and put Keren and Shannon in hammocks and hung the hammocks from logs they slung over their shoulders fore and aft and hauled them over to the Madeira.

A day and a night had gone by. On the Madeira, as muddy as the Mississippi and as wide at the mouth, they caught a ship with three decks, one above the other. It went up and down the river like a public bus. They had a three-day trip ahead of them...with no cabins or any other form of privacy except for a single bathroom on the first deck (for about two hundred passengers on a boat designed for ninety-nine, maximum) and no seats; instead, grossly over-crowded ranks of hammocks bearing a jam-up of people hanging shank to flank from the ceilings with their hummocky hips choking the air...

By now, Keren and Shannon were both suffering from severe diarrhea in addition to the fever and pain. Fortunately, Everett had brought along a chamber pot. Right there in the midst of the other passengers' hammock-swaddled bottoms, Keren and Shannon took turns sitting on the pot. Everett wrapped a blanket around each one like a tent with a head popping out at the top. The Brazilians couldn't keep their eyes off the gringos who were gushing gringo misery out of their hindsides. They were disgusted and riveted. They twisted over sideways in their hammocks so as not to miss a moment of the spectacle. The redheaded, redbearded gringo kept taking the pot of sloshing diarrheic rot through crowds of passengers, constantly bending way down with his reeking pot to pass under the hammocks or standing up with his reeking pot and leaning this way and that to hog-wrestle his way through the midair clutter of human haunches to reach the railing and dump the contents into the Madeira and weave his way back through the crowd with the chamber pot, knowing it would be no time before he had to slosh through them again with a potful of humiliation.

The spectators talked about them constantly, out loud and in full voice, apparently assuming that the gringos couldn't understand Portuguese. But Everett could.

"She's going to die, isn't she," one would cry, nodding toward Keren, who was down from 105 pounds to about 70, if that, and looked like the Red Death with a raging fever. "Of course she is," another would say. "Malaria does quick work with a skinnybones like that one."

Everett would experience a very small, rueful lift of superiority. These smug Brazilians obviously couldn't recognize typhoid fever when it was right in their faces.

People could already tell that Keren was dying! *One look* and they knew *that* much! Everett implored the captain, a one-armed Brazilian who was also the owner, to go faster, straight to Porto Velho. Skip the stops in between! My wife is *dying!*

"Look, comrade," said the one-armed Brazilian, without so much as a trace of fellow feeling, "if your wife is supposed to die, that's that. I won't speed up for you."

In what seemed like barely an hour the ship pulled into shore in the middle of nowhere and stopped. No passengers were getting on or off. There was no platform and barely a dock. Unaccountably, the entire crew had slipped on red jerseys, even the one-armed captain. With a *whoop* they all left the ship and climbed up a steep embankment. They looked like a lot of little ladybugs on the way up. At the top, men in green jerseys awaited them.

Godalmighty—they were stopping to play soccer!—and obviously they had arranged it well in advance.

Keren, her face a fiery red, slipped in and out of consciousness. It was two hours before the one-armed captain and his crew returned to the ship, still togged out in soccer gear, in high spirits, laughing, making jokes, jolly jolly jolly flirting with pretty girls among the passengers.

An eternity it took, but they finally reached the hospital in Porto Velho.

"My wife and my daughter have typhoid fever," Everett announced.

The doctor took a good look at Keren and Shannon and said, "Looks like malaria to me." He took drops of blood from Keren's and Shannon's fingers and put them on slides and examined them with a microscope... and began chuckling.

Indignant, Everett said, "What are you laughing at?"

"They have malaria, all right," he said, "and not just a little."

He laughed some more, apparently at Everett's ignorance. What made it even funnier was that Keren's and Shannon's bloodstreams had the highest levels of malaria he had ever come across in his whole career, and he treated malaria patients every day *hahahaaaa!*

Every doctor, every nurse, every AD (Almost a Doctor), told Everett that Shannon might make it, but it was too late for Keren. He had wasted so much time with his own AD diagnosis of typhoid fever, she would never survive.

But after two weeks of intensive care, she did—and would probably recover entirely...in time...which turned out to be six months' recuperation in her parents' home. Then she headed right back to the Pirahã and Everett.

Everett tells that story early in the book...then doesn't hesitate to turn to such matters as experiments on Pirahã numerosity, i.e., linguistic and psychological expression and control of numerical concepts. He weaves these dissertations throughout *Don't Sleep, There Are Snakes*...and it is hard to come away from the book without feeling they were just as important to him as the story of his life. And they probably were. They gave his saga some very necessary gravity...even as the story became more intense. The most intense was the night of the cachaça madness. This was three years into the Everetts' life among the Pirahã.

Cachaça is a liquor distilled from sugarcane. Brazilians had warned Everett about cachaça, but he had never actually had to deal with the problem before. Everett and his entire family—Keren, Shannon, Kristene, and Caleb—lived in one of the very few structures that rated the designation *house* in the Pirahã area of the jungles along the Maici River. It was built atop a four-foot-high platform. In the middle of the house was a storeroom. One night about nine o'clock the whole family was asleep when Everett heard loud talk and laughter on the riverbank. *Drunken* talk and laughter, if he knew anything about it. So he got up and went down to check it out. A boat such as Brazilian river traders use, a big one, had pulled into shore, and ten or twelve Pirahã were on the deck laughing and carrying on. They fell silent when they saw Everett approaching. There was no visual evidence of anybody drinking. So Everett settled for giving the captain, a Brazilian, a little lecture in Portuguese about how selling alcohol in this part of the Amazon was illegal and punishable by heavy fines and two years in jail. It occurred to him later that he must have sounded terribly officious, since technically he was nothing but a nosy American with a visa, commander of nothing. By the time he went back to bed, the noise had resumed; but he managed to fall asleep, only to be awakened an hour or so later by two men speaking in Pirahã inside a small house the Catholic missionary, by then departed, had built no more than a hundred feet away.

One Pirahã said, "I am not afraid. I kill the Americans. We kill them, the Brazilian gives us a new shotgun. He told me that."

"You kill them, then?" said the other.

"Yes. They go to sleep. I shoot them."

A bolt of panic through the solar plexus. Everett can tell they're merely waiting to work up the nerve or the cachaça blood level to do it. What earthly chance do he and Keren and the three children have? Exactly one, Everett concludes. He leaves the house immediately, as is, in his shorts and flip-flops. God, it's dark out here, blacker than black, and he didn't dare bring a flashlight because they might see him coming. Very odd!—no campfires such as the Pirahã keep lit out in front of their huts and lean-tos at night. (He would learn later that all the Pirahã women had put out the fires and fled deep into the jungle the moment they heard the word "cachaça.") Everett bursts into the drunk Pirahã's little hideaway next door with a big grin and, in Pirahã, gives them the merriest, liveliest "Hey guys! What's up!" any walking dead man ever exclaimed to his executioners. Without any pause at all he continues drenching them with the most hyperexuberant happy patter-blather imaginable, as if there had never been any closer comrades on this earth. *Oh, the times we've had together!* The drunk Pirahã stare at him without a word, utterly, boozily stupefied...as he gathers up all the weapons, two shotguns, two machetes, bows and arrows, and leaves chundering still more ebullient, chummy-honey-rummy talk all over them, flashing still more inexplicably ecstatic grins, even warbling bird words so sublimely that the most lyrical nightingale would exhale with hopeless envy—and once out the door and into the dark, cradling the cache of weaponry in his arms, he runs, actually *runs,* hobbledy, staggerly, stumbly, jackleggedly, back to his own house, where he stashes the cache in the storeroom, save for one shotgun. He removes the gun's shells. Then he has Keren and the children go into the storeroom and lock it from within. He remains on the platform just outside the door to the house and sits on a bench with the shotgun in his lap. Not even the most schnockered, hopelessly cachaça'd Pirahã can miss it.

He can hear Pirahã running toward him in the dark, ululating devastation. Other voices keep warning them,

"Watch out! Dan's got weapons!" ... weapons—*plural*—as if he's a one-Crooked-Head army. He's conscious of an arrow whizzing by, but not aimed at him. They don't dare shoot arrows at the one-Crooked-Head army. By 4:00 a.m. the Pirahã yahoos have headed down the riverbank, judging by the noise. Everett, exhausted, a nervous wreck, goes into the storeroom with everybody else, collapses asleep—

—*thump crash arrggggh* more *thuds groans* agonized *grunts*...on all sides of the storeroom only *one inch* away—that being the thickness of the storeroom wall right behind them, right before them, on both sides of them—*one inch* away—when *splat hock jaaaggh thump yaaak groan eeeowww* the whole storeroom shakes, shudders, trampolines on the platform...the bastards could break into this little room *just like that* if they knew we were in here, but now they're bent only upon beating each other senseless *thump crack groan ooof ummmph*. Everett has the shotgun with no shells to—what?—scare them with? There are so many Everetts young and old crammed in here, how can he possibly—

Gradually the fighting subsides...the house grows silent...they must have beaten each other into jelly by now...not a sound in the village, either...light visible through the storeroom's minute joined apertures...Everett dares open the hatchway...broad daylight...everywhere, an unnatural quiet...the whole village has sunk into a cachaça'd hangover.

Everett and Keren go through the house, expecting the worst. It was and it wasn't. There wasn't a whole lot of physical damage. Most of the damage the Pirahã had inflicted upon each other's cachaça'd selves. There were smears of blood everywhere—on the walls, on the beds…pools of blood on the floors…cachaça had turned their happy, laughing selves into blithering maniacs out for blood.

Later on, a ragtag delegation of Pirahã with black eyes, swollen jaws, and fat lips comes to Everett's house to apologize. They're friendly, indolent, loose, laid-back, lazy. Soon they will forget they went mad from cachaça: Who remembers "other day" anyway?

Don't Sleep, There Are Snakes instantly became a hit and the biggest wallop in the breadbasket Noam Chomsky's hegemony had ever suffered. Everett didn't so much attack Chomsky's theory as dismiss it. He spoke of Chomsky's "waning influence" and the mounting evidence that Chomsky was wrong when he called language "innate." Language had not evolved from…*anything*. It was an artifact. Just as man had taken natural materials, namely, wood and metal, and combined them to create the ax, he had taken natural sounds and put them together in the form of codes representing objects, actions, and, ultimately, thoughts and calculations—and called the codes *words*. In *Don't Sleep, There Are Snakes,* Everett animates his avant-garde theory with the story of his own thirty years with this, the most primit*—*er*, indigenous—tribe known to exist on earth, the Pirahã...risking death in virtually every conceivable form in the jungle, from malaria to murder to poison to getting swallowed by anacondas.

National Public Radio read great swaths of the book aloud over the their national network and named it one of the best books of the year.[2] Reviews in the popular press were uniformly favorable, even glowing...to the point of blinding...as in the *Sacramento Book Review:* "A genuine and engrossing book that is both sharp and intuitive; it closes around you and reaches inside you, controlling your every thought and movement as you read it." It is "impossible to forget."[3]

Ideally, great wide-eyed romantic acclaim like this should have no effect, except perhaps a negative one, in academia. But when the truth squad's forty-thousand-word "reassessment" finally came out in *Language,* in June of 2009, there was no explosion. The Great Rebuttal just lay there, a swollen corpus of objections—cosmic, small-minded, and everything in between. It didn't make a sound. The success of *Don't Sleep, There Are Snakes* had defused it.

Chomsky and the squad were far from done for, however.

---

* According to Merriam-Webster's online dictionary, the word "primitive" can be defined as: "of, belonging to, or seeming to come from an early time in the very ancient past; not having a written language, advanced technology, etc.;...of, relating to, or produced by a people or culture that is nonindustrial and often nonliterate and tribal."

They concentrated on the academic press. No academic, in what was still the Age of Chomsky, was likely to write any gushing review of Everett's scarlet book. Chomsky and the squad were on the qui vive for anyone who stepped out of line. A professor of philosophy at King's College London, David Papineau, wrote a more or less positive review of *Don't Sleep*—only that: "more or less"—and a member of the truth squad, David Pesetsky, put him in his place. Papineau didn't take this as good-hearted collegial advice. "For people outside of linguistics," he said, "it's rather surprising to find this kind of protection of orthodoxy."[4]

Three months after *Don't Sleep* was published, Chomsky dismissed Everett to the outer darkness with one of his favorite epithets. In an interview with *Folha de S.Paulo,* Brazil's biggest and most influential newspaper, news website, and mobile news service, Chomsky said Everett "has turned into a charlatan."[5] A charlatan is a fraud who specializes in showing off knowledge he doesn't have. The epithets ("fraud," "liar," "charlatan") were Chomsky's way of sentencing opponents to Oblivion. From then on Everett wouldn't rate the effort it would take to denounce him.

Everett had, as it says in the song, let the dogs out. Linguists who had kept their doubts and grumbles to themselves were emboldened to speak out openly.

Michael Tomasello, a psychologist who was codirector of the Max Planck Institute for Evolutionary Anthropology and one of the scholars who commented on Everett's 2005 article in *Current Anthropology,* had been critical of this and that in Chomsky's theory for several years. But in 2009, after Everett's book was published, he went all out in a paper entitled "Universal Grammar Is Dead" for the journal *Behavioral and Brain Sciences* and confronted Chomsky head-on: "The idea of a biologically evolved, universal grammar with linguistic content is a myth."[6] "Myth" became the new word. Vyvyan Evans of Wales's Bangor University expanded it into a book, *The Language Myth,* in 2014. He came right out and rejected Chomsky's and Steven Pinker's idea of an innate, natural-born "language instinct." In a blurb, Michael Fortescue of the University of Copenhagen added, "Evans' rebuttal of Chomsky's Universal Grammar from the perspective of Cognitive Linguistics provides an excellent antidote to popular textbooks where it is assumed that the Chomskyan approach to linguistic theory…has somehow been vindicated once and for all."[7]

Thanks to Everett, linguists were beginning to breathe life into the words of the anti-Chomskyans of the twentieth century who had been written off as cranks or contrarians, such as Larry Trask, a linguist at England's University of Sussex. In 2003, the year after Chomsky announced his Law of Recursion, Trask said in an interview, "I have no time for Chomskyan theorizing and its associated dogmas of 'universal grammar.' This stuff is so much half-baked twaddle, more akin to a religious movement than to a scholarly enterprise. I am confident that our successors will look back on UG as a huge waste of time. I deeply regret the fact that this sludge attracts so much attention outside linguistics, so much so that many non-linguists believe that Chomskyan theory simply is linguistics...and that UG is now an established piece of truth, beyond criticism or discussion. The truth is entirely otherwise."[8]

In 2012 Everett published *Language: The Cultural Tool,* a book spelling out in scholarly detail the linguistic material he had tucked in amid the tales of death-dodging in *Don't Sleep, There Are Snakes*...namely, that speech, language, is not something that had *evolved* in *Homo sapiens,* the way the breed's unique small-motor-skilled hands had...or its next-to-hairless body. Speech is man-made. It is an artifact...and it explains man's power over all other creatures in a way Evolution all by itself can't begin to.

Language: The Cultural Tool was Everett's *Origin of Species,* his *Philosophiae Naturalis*...and it wasn't nearly the success that *Don't Sleep* had been. It went light on the autobiographical storytelling...Oh, the book had its moments...Only Everett had it in him to make direct fun of Chomsky...He tells a story about visiting MIT in the early 1990s and going to what was billed as a major Chomsky lecture. "A group of his students were sitting in the back giggling," says Everett. "When Chomsky mentioned the Martian linguist example, they could barely constrain their chuckles and I saw money changing hands." After the talk, he asked them what that was all about, and they said they had bets with each other on exactly when in his lecture Chomsky would drop his moldy old Martian linguist on everybody.

Critics such as Tomasello and Vyvyan Evans, as well as Everett, had begun to have their doubts about Chomsky's UG. Where did that leave the rest of his anatomy of speech? After all, he was very firm in his insistence that it was a physical structure. Somewhere in the brain the *language organ* was actually pumping the UG through the *deep structure* so that the LAD, *the language acquisition device,* could make language, speech, audible, visible, the absolutely real product of *Homo sapiens*'s central nervous system.

And Chomsky's reaction? As always, Chomsky proved to be unbeatable when it came to debate. He never let himself be backed into a corner, where he could be forced to have it out with his attackers jowl to howl. He either jumped out ahead of them and up above them or so artfully dodged them that they were left staggering off stride. Tomasello had closed in and just about *had* him on all this para-anatomy, when suddenly—

—*shazzzzammm*—Chomsky's language organ and all its para-anatomy, if that was what it was, disappeared, as if it had never been there in the first place. He never recanted a word. He merely subsumed the same concepts beneath a new and broader body of thought. Gone, too, astonishingly, was recursion. *Recursion!* In 2002 Chomsky had announced his discovery of recursion and pronounced it *the* essential element of human speech. But here, in the summer of 2013, when he appeared before the Linguistic Society of America's Linguistic Institute at the University of Michigan...recursion had vanished, too. So where did that leave Everett and his remarks on recursion? Where? Nowhere. Recursion was no longer an issue...and Everett didn't exist anymore. He was a ghost, a vaporized nonperson. Naturally, the truth squad could no longer see him, either. They couldn't have cared less about churning up an angry wave for *Language: The Cultural Tool* to come surfing in on. They didn't even extend Everett the courtesy of loathing him in print. They left non-him behind with all the rest of history's roadside trash.

The passage of time did not mollify Chomsky's opinion of the non-him, Everett, in the slightest. In 2016, when I pressed him on the point, Chomsky blew off Everett like a nonentity to the minus-second power.

"It"—Everett's opinion; he does not refer to Everett by name—

> amounts to absolutely nothing, which is why linguists pay no attention to it. He claims, probably incorrectly...it doesn't matter whether the facts are right or not. I mean, even accepting his claims about the language in question—Pirahã—tells us nothing about these topics. The speakers of this language, Pirahã speakers, easily learn Portuguese, which has all

the properties of normal languages, and they learn it just as easily as any other child does, which means they have the same language capacity as anyone else does. Now, it's conceivable, though unlikely, that they just don't bother using that capacity. It's like finding some kind of bird that could fly freely but just doesn't bother going up above trees. I mean, it's conceivable, pretty unlikely but conceivable. And it would tell you nothing about biology.[9]

As a result, Everett's new book didn't begin to kick up the ruckus that *Don't Sleep, There Are Snakes* had. An entirely new world had been born in linguistics. In effect, Chomsky was announcing—without so much as a quick look back over his shoulder—"Welcome to the Strong Minimalist Thesis, Hierarchically Structured Expression, and Merge." A regular syllablavalanche had buried the language organ and the body parts that came with it.

Starting in the 1950s, said Chomsky, whose own career had started in the 1950s, "there's been a huge explosion of inquiry into language.... Far more penetrating work is going on into a vastly greater array of theoretical issues.... Many new topics have been opened. The questions that students are working on today could not even be formulated or even imagined half a century ago or, for that matter, much more recently...." They are "considering more seriously the most fundamental question about language, namely, what is it."

What is it?! With the help of "the formal sciences," said Chomsky, we can take on "the most basic property of language, namely, that each language provides an unbounded array" of (Chomsky loved "array") "hierarchically structured expressions...through some rather obscure system of thought that we know is there but we don't know much about it."[10]

The following year, in August of 2014, Chomsky teamed up with three colleagues at MIT, Johan J. Bolhuis, Robert C. Berwick, and Ian Tattersall, to publish an article for the journal *PLoS Biology* with the title "How Could Language Have Evolved?" After an invocation of the Strong Minimalist Thesis and the Hierarchical Syntactic Structure, Chomsky and his new trio declare, "It is uncontroversial that language has evolved, just like any other trait of living organisms." Nothing else in the article is anywhere nearly so set in concrete. Chomsky *et alii* note it was commonly assumed that language was created primarily for communication...*but*...in fact communication is an all but irrelevant, by-the-way use of language...language is deeper than that; it is a "particular computational cognitive system, implemented neurally"...*but*..."we are not sure exactly how"...there is the proposition that Neanderthals could speak...*but*...there is no proof...we know anatomically that the Neanderthals' hyoid bone in the throat, essential for *Homo sapiens*'s speech, was in the right place... *but*... "hyoid morphology, like most other lines of evidence, is evidently no silver bullet for determining when human language originated"...Chomsky and the trio go over aspect after aspect of language...*but*...there is something wrong with every hypothesis...they try to be all-encompassing...*but*...in the end any attentive soul reading it realizes that all five thousand words were summed up in the very first eleven words of the piece, which read:

"The evolution of the faculty of language largely remains an enigma."

An enigma! A century and a half's worth of certified wise men, if we make Darwin the starting point—or of bearers of doctoral degrees, in any case—six generations of them had devoted their careers to explaining exactly what language is. After all that time and cerebration they had arrived at a conclusion: language is...*an enigma?* Chomsky all by himself had spent sixty years on the subject. He had convinced not only academia but also an awed public that he had the answer. And now he was a signatory of a declaration that language remains...*an enigma?*

"Little enough is known about cognitive systems and their neurological basis," Chomsky had said to John Gliedman back in 1983. "But it does seem that the representation and use of language involve specific neural structures, though their nature is not well understood."

It was just a matter of time, he intimated then, until empirical research would substantiate his analogies. That was thirty years ago. So in thirty years, Chomsky had advanced from "specific neural structures, though their nature is not well understood" to "some rather obscure system of thought that we know is there but we don't know much about it."

In three decades nobody had turned up any hard evidence to support Chomsky's conviction that every person is born with an innate, gene-driven power of speech with the motor running. But so what? Chomsky had made the most ambitious attempt since Aristotle's in 350 BCE to explain what exactly language is. And no one else in human history had come even close. It was dazzling in its own flailing way—this age-old, unending, utter, ultimate, universal display of ignorance concerning man's most important single gift.

Language—what is it? What is it? Chomsky's own words at age eighty-five after a lifetime of studying language! The previous 150 years had proved to be the greatest era ever in solving the riddles of *Homo sapiens*—but not in the case of *Homo loquax,* man speaking. A parade of certified geniuses had spent lifetimes trying to figure it out—and failed.

The first breakthrough, leading at last to the answer, was Everett's thirty-year study of the Pirahã in their remote and forgotten malarial jungle. Historians often idly wish they could somehow spend just a little time, even fifteen minutes, in the worlds they're writing about. In effect, that was what Everett was doing...when he converted from

his Christian faith to linguistics and anthropology. The Pirahã were not frozen in time. They were living in real time
and using man's greatest artifact, language, as best they
could...with the world's smallest vocabulary. They had no
one to pressure them, cajole them, or force them—say,
through military coercion—to change. Without planning it,
Everett found himself studying a language not by dissect-
ing and analyzing it as a finished product, the way it existed
in Europe, the United States, or the Pacific Rim, but by
starting with a prototype. And in the Pirahã he had found
the most basic prototype of *Homo sapiens*. The Pirahã lived
entirely in the present, spoke only in the present tense,
did not analyze their past or agonize over their future—
which in no small part accounted for their generally ami-
able, relaxed, laugh-light demeanor. For the same reason,
they spent virtually no time making artifacts, not even the
simplest. Artifacts are an elementary part of thinking about
the future.

Most tellingly of all, they had no social gradations, no hierarchy of social classes, and not even any status groups, so
far as Everett had been able to determine. They didn't have
occupations...wise men here, fighting men there, spokes-
men, repairmen, builders, messengers, young rips, party
girls, or, for that matter, parties.

In 1869, under the pressure of competition with Wallace, Darwin had come up with the theory that speech
evolved from man's imitation of birdsong. As the sounds

became more complex, he explained in 1871 in *The Descent of Man,* they developed into what we now know to be words... Darwin said that?... Birdsong?... It was not a very convincing notion, and a year later the seventy-seven-year blackout began. When it ended after World War II, linguists, philosophers, psychologists, and even paleontologists began chundering out a vast Cloud of theories of the evolution of language as if to make up for lost time.

Scientificalization was the intellectual spirit of the age, and nobody filled that bill better than Noam Chomsky. The origin of language, he theorized, was a chance mutation that occurred in a single individual and evolved into its finished and entirely physical form in less than two hundred thousand years... a mere blink in the conventional chronology of Darwinian evolution.

Michael Tomasello put forward his gestural theory. It held that once man evolved to standing on two legs, his hands were free to transmit signals... until eventually the signals, the gestures, evolved into speech.

Both W. Tecumseh Fitch and Kenneth Kaye believed that the sounds mothers cooed, crooned... or growled... to their babies, the "motherese," had evolved into words.

Erich Jarvis of Duke University formed a team that picked up where Darwin left off. They isolated forty-eight bird genomes, made a digital slew of studies of birds' vocal learning, had a "computational biologist" crunch the numbers—and found that "the same genes that give humans the ability to speak give birds the ability to sing."[11] Ryuji Suzuki of the Howard Hughes Medical Institute and MIT led a team that developed a computer algorithm to analyze the moans, cries, and chirps in sixteen different love songs, from six to thirty minutes each, that male humpback whales croon during the mating season. These great blubbery beasts use a hierarchical syntax similar to man's, although the gigabytten Suzuki was the first to admit that he hadn't the faintest notion what they were singing.

Johan J. Bolhuis headed a team of five neuroscientists who studied a musical bird, the zebra finch, by playing adult zebra finches tapes of songs they had listened to as little birds in the wild... and now that they were mature, analyzing their sexual arousal (somehow) when they heard them again.

Three Japanese psychologists plus one American, Robert C. Berwick, came up with the Integration Hypothesis... which says that human language has two components: E for "expressive," as in birdsong, and L for "lexical," as in monkey cries. In man, E and L come together to create human language. Why were they eager to have Berwick on the team? Because he was an MIT "computational linguistics" star who knew how to parameterize—that's the word, *param*eterize—any linguistic theory into modules and press a button and run them through his Prolog system and *just like that* determine how the theory works for any or all of several dozen languages in terms of "psycholinguistic fidelity" and "logical adequacy."

The math-lang specialists at Princeton's Institute for Advanced Study, Martin Nowak and David Krakauer, wrote, or, rather, mathed up, an article for *Proceedings of the National Academy of Sciences* entitled "The Evolution of Language"—in terms of so-called evolutionary game theory. Game theory produces articles in which equations bear such heavy loads of calculus that the linguists who come along as their divan carriers buckle under the weight—and Nowak and Krakauer were heavyweights. Their equations made Swadesh's glottolingo back in the late 1940s and 1950s look absolutely limpid by comparison. At one point, they calculate the effect that "errors in perception" must have had early in the evolution of language. "Signals are likely to have been noisy and can therefore be mistaken for each other. We denote the probability of interpreting sound i as sound j by u_{ij}. The payoff for L communicating with L' is now given by

$$F(L, L') = \frac{1}{2} \sum_{i=1}^{n} \sum_{j=1}^{n} \left[p_{ij} \left(\sum_{k=1}^{m} u'_{jk} q'_{ki} \right) + p'_{ij} \left(\sum_{k=1}^{m} u_{jk} q_{ki} \right) \right]$$

It was the greatest array—yes! *array!*—of mathematical high fliers ever assembled to go where no man has gone before…up into the digital Cloud to solve the mystery of language—

155

—and all ran smack into a firewall and came no closer to finding the answer to *Language—what is it?* than the rest. It was a long, dreary stretch of failure. Linguists, philosophers, anthropologists, and psychologists had been trying on and off forever to figure out exactly what language was and were no closer to solving the mystery than Darwin had been with his songbirds. A century and a half... and *nothing* to show for it. In May of 2014, Chomsky, Tattersall, Berwick the data cruncher, Marc Hauser of the 2002 Recursion Three, plus four other eminenti—making eight in all—published a historic ten-thousand-word revelation entitled "The Mystery of Language Evolution." Historic it was, but not in any triumphant sense. In fact, there had never been a scholarly paper quite like it. Here you had a delegation of some of the biggest names in the study of language, above all, Chomsky, running up a white flag of abject defeat and surrender... after forty straight years of failure.

"In the last 40 years," this eight-man jeremiad began—as we first heard on the opening pages of this book—"there has been an explosion of research on this problem as well as a sense that considerable progress has been made. We argue instead that the richness of ideas is accompanied by a poverty of evidence, with essentially no explanation of how and why our linguistic computations and representations evolved."[12] There is nothing whatsoever like it in animal life, the Eight continued. Fossils and archaeology tell us nothing. No one has been able to find any genetic roots of language. There are no empirical tests of any hypotheses. "The most fundamental questions about the origins and evolution of [language] remain as mysterious as ever," and we don't know if we will ever be able to find a way to answer them. The Eight were neo-Darwinists to a man, hard-core Evolutionists who still believed in Dobzhansky's maxim from the 1930s: "Nothing in biology makes sense except in the light of evolution." Did their capitulation lead them to turn to Everett and his idea that language just might be an artifact in and of itself, every bit as much as a lightbulb or a Buick is? Not for a moment. Linguists who had begun to forsake Chomsky's notion of a language organ—and Chomsky had as much as forsaken it himself—did not care to go in that direction. Everett was still the flycatcher non grata. Those who admitted that well, yes, language did not evolve in the usual way tended to depict language as one of Evolution's fellow travelers. The favorite new phrase was "biological niche construction." The niche was hollowed out into Evolution's flanks for the long march. It never seemed to dawn on them that they were indulging in sheer metaphor.

In 2015 a leading niche man, Chris Sinha of China's Hunan University, wrote that "niche construction theory is a relatively new approach in evolutionary biology that seeks to integrate an ecological dimension into the Darwinian Theory of Evolution by natural selection."[13] In this case "ecological" meant something picked up in the course of Evolution's interaction with the environment. This and compromise theories like it left only a scattering of subscribers to the idea that language is an out-and-out artifact...notably, Edward Sapir back in the 1920s, before his theory was trampled in the rush toward Chomskyism...Andy Clark, then of Washington University in Saint Louis, a philosopher who in 1997 called language "the ultimate artifact"[14]...and Everett. Everett's decades-long experience with the Pirahã cast the first light on the answer to Chomsky's question after sixty years of studying language: *What is it?*

Everett didn't know philosopher Andy Clark or his work. He arrived at Clark's insight on his own: language is the great "cultural tool," as Everett called it. Everett never showed any sign of doubting the Theory of Evolution. Why should he? Speech, language, was something that existed quite apart from Evolution. It had nothing whatsoever to do with it. Man, man unaided, created language. Everett's leap over the wall Darwin had built around Evolution—namely, Darwin's conviction that the Theory of Evolution was also a Theory of Everything—Everett's leap over *that* wall was one that next to no Evolutionist had even contemplated.

By now, 2014, Evolution was more than a theory. It had become embedded in the very anatomy, the very central nervous system, of all *modern* people. Every part, every tendency, of every living creature had evolved from some earlier form—even if you had to go all the way back to Darwin's "four or five cells floating in a warm pool somewhere" to find it. A title like "The Mystery of Language Evolution" was instinctive. It went without saying that any "trait" as important as speech had *evolved*...from *something*. Everett's notion that speech had not evolved from *anything*—it was a "cultural tool" man had made for himself—was unthinkable to the vast majority of *modern* people. They had all been so deep-steeped in the Theory that anyone casting doubt upon it obviously had the mentality of a Flat Earther or a Methodist. With the very title of his book itself—*Language: The Cultural Tool*—Everett was drawing the same line Max Müller had drawn in 1861: "Language is our Rubicon, and no brute will dare to cross it."

Whether it was Darwin alone at his desk in 1870 thinking, between bouts of vomiting, of ways to get around Alfred Wallace's objections that he couldn't account for speech...or Chomsky plus the truth squad with Berwick at the keyboard chundering out fantastical arrays of calculus Σ sigma Σ blades...or any of the scores of hypotheses in the 150 years in between...all were based on the same "uncontroversial" assumption. That was Chomsky's and his MIT colleagues' very word in their August 2014 article in *PLoS Biology* entitled "How Could Language Have Evolved?" That it had evolved was a given. Rare was the linguist, the psychologist, the anthropologist who could entertain the notion that something as fundamental to human life as language might be an artifact.

The answer was to come not from the digital universe...but from analogical terrain that seemed sunk in the past, not only because the concept is so simple that at first most linguists couldn't think of it as a concept. It comes down to a single word: mnemonics

By 350 BCE both Plato and Aristotle were writing about mnemonics, and Aristotle was working on a complete system of analysis. The word "mnemonic" is derived from the Greek *mnemom,* meaning "mindful." The *m* is silent, like the *p* in "pneumonia." A mnemonic is a device, essentially a trick, a sleight-of-mind, an easy-to-remember key for opening up a body of knowledge too long, too detailed, too cumbersome, too complicated, or simply too tiresome, too annoying, to have to memorize without a memory aid—which is the two-word definition of a mnemonic. The Greeks favored so-called topical, or loci (meaning "locations"), mnemonics, in which each term, name, or number is imagined to be in a particular area of a certain room on a certain floor in one of an endless row of identical houses. It was surprisingly easy to gather them up later in the right order. Probably the best known mnemonics in English are "metrical" mnemonics: "Thirty days hath September, April, June, and November"..."*I* before *E* except after *C*"..."In fourteen hundred and ninety-two Columbus sailed the ocean blue"..."Red sky at night, shepherd's delight; red sky in morning, shepherd's warning."

Today mnemonics is not thought of as anything more

practical than a memory device for remembering ingredients, lists, and in some cases formulas. Virtually all the sciences depend upon mnemonics, typically in the form of sentences or phrases in which the first letter of each word stands for a different item or procedure—or even in the form of a single word whose letters stand for different components. Chemistry, for example, produces mnemonics by the gross. Some are rather clever, such as the one for organic chemistry's sequence of dicarboxylic acids: oxalic, malonic, succinic, glutaric, adipic, pimelic, suberic, azelaic, and sebacic. If you capitalize the first letter of each word and are clever enough, you can come up with *"Oh My, Such Good Apple Pie, Sweet As Sugar."* That's the mnemonic.

The sequence of orbitals (areas in which electrons move), designated **s p d f g h i k,** is transformed into a mnemonic almost as easy to remember: *Sober Physicists Don't Find Giraffes Hiding In Kitchens.*

The lanthanide series of elements, which goes "La Ce Pr Nd Pm Sm Eu Gd Tb Dy Ho Er Tm Yb Lu," generates *"Lame Celibate Prudes Need Promiscuous Smut with European Gods. Troublesome Dying Ho, it's Erotic TiMe, You Bitch LUst!"*

One of the so-called activity series of metals goes "K>Na>Mg>Al>Zn>Fe>Pd>H>Cu>Au" and gets neatly mnemonicked into *"Kangaroos Naturally Muck About in Zoos For Purple Hippos Chasing Aardvarks."*

And no student of the mnemonic arts is likely to leave out

the one that turns the mile-long electrochemical series "potassium>sodium>calcium>magnesium>aluminum>zinc>iron>tin>lead>hydrogen>copper>silver>gold" into a mini poem: *"Paddy Still Could Marry A Zulu In The Lovely Honolulu, Causing Strange Gazes."*

Language itself, the mother of all mnemonics, is precisely the same sort of device that chemistry employs. Words are elemental mnemonics, sequences of sounds (the alphabet) used to remember everything in the world, from the smallest to the greatest. Speech, language, is a matter of using these mnemonics, i.e., words, to create meaning.

And that is all that speech is, a mnemonic system—one that has enabled *Homo sapiens* to take control of the entire world. It is language, and only language and its mnemonics, that creates memory as *Homo sapiens* experiences it. Even the smartest apes don't have thoughts so much as conditioned responses to certain primal pressures, chiefly, the need for food and the fear of physical threats.

But in point of fact mnemonics isn't just in the service of language. Mnemonics *is* language. Throughout the history of language—and it's quite irrelevant to try to make the usual paleontological guesses as to when that was—man has converted objects, actions, thoughts, concepts, and emotions into codes, conventionally known as words. No one now knows...and there is no reason why anyone is likely to ever know...when it occurred to *Homo sapiens* to use words as mnemonics. But there are now between six thousand and seven thousand different mnemonic systems, better known as languages, covering the world today. They, and they alone, *are* language...they are simple and clear. Perhaps it can be amusing to watch otherwise bright people banging their brainpans into the same firewall, herds of them, schools of them, generations of them, whole Eras and Ages of them, an entire bright universe of them, endlessly—but for how long?

Bango! One bright night it dawned on me—not as a profound revelation, not as any sort of analysis at all, but as something so perfectly obvious, I could hardly believe that no licensed savant had ever pointed it out before. There *is* a cardinal distinction between man and animal, a sheerly dividing line as abrupt and immovable as a cliff: namely, speech.

"Speech," I said to myself, "gave the human beast far more than an ingenious tool for communication. Speech was a veritable nuclear weapon!"

Speech was the first artifact, the first instance in which a creature, man, had removed elements from nature...in this case, sounds...and turned them into something entirely new and man-made...strings of sounds that formed codes, codes called words. Not only is speech an artifact, it is the primal artifact. Without speech the human beast couldn't have created any other artifacts, not the crudest club or the simplest hoe, not the wheel or the Atlas rocket, not dance, not music, not even hummed tunes, in fact not tunes at all, not even drumbeats, not rhythm of any kind, not even keeping time with his hands.

Speech, and only speech, gives the human beast the ability to make plans...not just long-term but *any* plans, even for something to do five minutes from now. Speech, and only speech, gives the human beast the power of accurate memory and the means to preserve it in his thoughts for now or indefinitely in print, in photographs, on film, or in the form of engineering and architectural diagrams. Speech, and only speech, enables man to use mathematics. (Doubters need only try to count from one to ten without words.) Speech, and only speech, gives the human beast the power to enlarge his food supply through an artifice called farming. Speech ended not only the evolution of man, by making it no longer necessary for survival, but also the evolution of animals.

Today the so-called animal kingdom is an animal colony, and we own it. It exists only at our sufferance. If we were foolish enough and could get the cooperation of people all over the earth, in six months we could exterminate every animal that sticks up more than a half inch above the ground. Already all cattle, chickens, and sheep in the world and the vast majority of pigs, horses, and turkeys—we hold the whole huge gaggle of them captive, *all* of them...to do with as we wish.

In short, speech, and only speech, has enabled us, we human beasts, to conquer every square inch of land in the world, subjugate every creature big enough to lay eyes on, and eat up half the population of the sea.

And this, the power to conquer the entire planet for our own species, is the minor achievement of speech's great might. The great achievement has been the creation of an internal self, an *ego*. Speech, and only speech, gives man the power to ask questions about his own life—and take his own life. No animal ever commits suicide. Speech, and only speech, gives us the urge to kill others on a massive scale, whether in war or other campaigns of terror. Speech, and only speech, gives us the power to exterminate ourselves and render the planet uninhabitable *just like that* in a matter of thirty-five or forty nuclear minutes. Only speech gives man the power to dream up religions and gods to animate them...and in six extraordinary cases to change history—for centuries—with words alone, without money or political backing. The names of the six are Jesus, Muhammad (whose military power came only after twenty years of preaching), John Calvin, Marx, Freud—and Darwin. And this, rather than any theory, is what makes Darwin the monumental figure that he is.

The human beast does not require that the explanation offer hope. He will believe whatever is convincing. Jesus offered great hope. The last shall be first and the first shall be last. It is easier for a camel to pass through the eye of a needle than for a rich man to enter the Kingdom of Heaven. The meek shall inherit the earth and ascend to the right hand of God. This, from the Sermon on the Mount, is the most radical social and political doctrine ever promulgated. Its soldiers were thousands, millions, of the meek, and it took the better part of three centuries for the Word to build up such a following that the Roman emperor Constantine converted to Christianity. Calvin offered less hope than Jesus; Muhammad, more and less; Marx, more and more. The meek—"the proletariat" he called them—shall inherit the earth *now!...here!...* and never mind waiting for Heavenly pie in the sky. Freud offered more sex. Darwin offered nothing at all. Each, however, has left an enduring influence.

Jesus is the Rock of Ages for both Marxism and its less vulgar child, Political Correctness in American colleges and universities, today, even though Jesus's latter-day ducklings would gag on the very thought. There was a seventy-two-year-old field experiment in Marxism, 1917–1989, that failed gruesomely. But Marx's idea of one social class dominating another may remain with us forever. In medical terms, Freud is now considered an utter quack and a dotty old professor. But his notion of sex as an energy like the steam in a boiler, which must be released in an orderly fashion or the boiler will blow up, remains with us, too. At this moment, as you gaze upon these pages, you can be sure that there are literally millions of loin spasms and convulsions taking place throughout the world that would not be occurring were it not for the words of Sigmund Freud.

And this, the power of one person to control millions of his fellow humans—for centuries—is a power the Theory of Evolution cannot even begin to account for…or abide. Muhammad's words have enthralled and ruled the daily lives of 35 percent of the people on earth since the eighth century. And that rule has only grown stronger in our time. Jesus's words held sway over a comparable percentage of the world's population for one and a half millennia before weakening in Europe during the last half of the twentieth century.

Words are artifacts, and until man had speech, he couldn't create any other artifacts, whether it was a slingshot or an iPhone or the tango. But speech, the font of all artifacts, had a life no other artifact would ever come close to. You could lay aside a slingshot or an iPhone and forget about it. You could stop dancing the tango and it would vanish forever…or until you deigned to dance again. But you couldn't make speech lie down once it left your lips. The same remark could make your nieces and nephews crack up with mirth and laughter and make your brothers and sisters loathe you forever. Mighty men could say the wrong thing, and tens of thousands of little men might lose their lives in the war that followed right after the words came out of his mouth. Or a weak man might get drunk one night and say something romantic to a pretty girl. He wakes up in the morning with a terrible hangover, kneading his forehead and consumed with guilt because of the sweet possessive looks she's giving him. She has no trouble putting him in a box and tying it with a ribbon and giving him to herself as a wedding gift...the kickoff of sixty-two years during which he has a chance to find out just how stupid she is and how lovely she isn't—all of it the result of a little drunk speech he uttered back in another century.

Soon speech will be recognized as the Fourth Kingdom of Earth. We have *regnum animalia, regnum vegetabile, regnum lapideum* (animal, vegetable, mineral)—and now *regnum loquax,* the kingdom of speech, inhabited solely by *Homo loquax.* Or is "kingdom" too small a word for the eminence of speech, which can do whatever it feels like doing with the other three—physically and in every other way? Should it be Imperium loquax, making speech an empire the equal of Imperium naturae, the empire of Nature? Or Universum loquax, the Spoken Universe...this "superior intelligence," this "new power of a definite character"?

Last night I was riffling through the pages of a textbook on Evolution. I came upon a two-page spread with a picture on the left-hand page of a chimpanzee and her baby settling in for the night upon a three-pronged fork in a tree. On the right-hand page was a picture of a troop of gorillas stamping down a stretch of underbrush into crude nests for the night.

I looked up from the book and out the window upon two rather swell hotels, just a few blocks from where I live in New York City, the Mark and the Carlyle, which is thirty-five stories high…two air-conditioned, centrally heated, room-serviced, DUX-mattressed, turned-down-quilted, down-lighter-lit, Wi-Fi-wired, flat-screen-the-size-of-Colorado'd, two-basin-bathroomed, debouched-silk-draped, combination-safed, School-of-David-Hicks-carpeted, Bose-Sound-systematic, German-brass-fixture-showered hotels…full of God knows how many humans who expect at least that much for their $750 per night and up…and in the distance the peaks of the Chrysler Building, the Empire State Building, the Citicorp Building, and the very tip of the top of the new Freedom Tower…and in between, a steel field of towers 10, 20, 30, 40, 50 stories high.

It occurred to me that the two bedtime scenes, Apeland's on the one hand and Manhattan's on the other, were a perfect graph of what speech hath wrought. *Speech!* To say that animals evolved into man is like saying that Carrara marble evolved in to Michelangelo's *David*. Speech is what man pays homage to in every moment he can imagine.

NOTES

CHAPTER I: THE BEAST WHO TALKED

1   Thomas Malthus, *An Essay on the Principle of Population* (London: J. Johnson, 1798), chapter 2.

2   Ibid., chapter 1.

3   James Hutton, *An Investigation of the Principles of Knowledge, and of the Progress of Reason, from Sense to Science and Philosophy,* (Edinburgh: Strahan and Cadell, 1794).

4   Erasmus Darwin, *Zoonomia; or, the Laws of Organic Life,* (London: J. Johnson, 1794).

5   Jean-Baptiste Lamarck, *Recherches sur l'organisation des corps vivants* (Paris: Maillard, 1802). The book was based on Lamarck's 1800 lecture at Paris's National Museum of Natural History, where he was a professor.

6   James Secord, *Victorian Sensation: The Extraordinary Publication, Reception, and Secret Authorship of "Vestiges of the*

170

Natural History of Creation" (Chicago: University of Chicago Press, 2003), 20–21.

7   Adam Sedgwick, "Vestiges of the Natural History of Creation," *Edinburgh Review* (July 1845), 1–85.

8   Adam Sedgwick to Charles Lyell, April 9, 1845, in *The Life and Letters of the Reverend Adam Sedgwick,* ed. John Willis Clark and Thomas McKenny Hughes (London: C. J. Clay and Sons, 1890), 83.

9   Quoted in John M. Lynch, ed., *Selected Periodical Reviews, 1844–54,* vol. 1 of *"Vestiges" and the Debate Before Darwin* (Bristol, UK: Thoemmes Press, 2000).

10   Ibid., 10.

11   Ben Waggoner, "Robert Chambers," University of California Museum of Paleontology, http://www.ucmp.berkeley.edu/history/chambers.html.

12   Thomas Henry Huxley, review of *Vestiges of the Natural History of Creation,* 10th ed., *The British and Foreign Medico-Chirurgical Review* 13 (January–April 1854), 438.

13   Ibid., 427.

14   Alfred Russel Wallace, *My Life: A Record of Events and Opinions* (London: Chapman & Hall, 1905), 361–63.

15   Alfred Russel Wallace, "On the Tendency of Varieties to Depart Indefinitely from the Original Type," *Journal of the Proceedings of the Linnean Society: Zoology* 3, no. 9 (August 20, 1858). Initially read at the July 1, 1858, meeting of the Linnean Society.

16   Darwin writes that Lyell had praised Wallace's work in a letter to Wallace dated December 22, 1857, https://www.darwinproject.ac.uk/letter/DCPLETT-2192.xml.

17 Mark Rothery, "The Wealth of the English Landed Gentry, 1870–1935," *Agricultural History Review* 55, no. 2 (2007), 251–68.

18 A record of Charles Lyell I purchasing the estate was published in the December 9, 1887, edition of the *Scottish Law Reporter,* which included his and his successors' professions.

19 See the Darwin family tree prepared by Charles Darwin in his book *The Life of Erasmus Darwin,* ed. Desmond King-Hele (New York: Cambridge University Press, 2004), 141–143.

20 Ibid., 25.

21 Michael Shermer, *In Darwin's Shadow: The Life and Science of Alfred Russel Wallace; A Biographical Study on the Psychology of History* (New York: Oxford University Press, 2002), 14–15.

22 Charles Darwin, *Narrative of the Surveying Voyages of His Majesty's Ships* Adventure *and* Beagle, *1832–1836,* vol. 3 (London: Henry Colburn, 1839), 95–96.

23 These stories have been communicated orally for generations, so there are no definitive versions. For more on these creation myths and others, see: www.powhatanmuseum.com.

24 George Thornton Emmons, *The Tlingit Indians,* ed. Frederica de Laguna (Seattle: University of Washington Press, 1991), 34.

25 David Adams Leeming, *A Dictionary of Creation Myths* (New York: Oxford University Press, 1996), 44–46.

26 Ibid., 252–53.

27 George Hart, *The Routledge Dictionary of Egyptian Gods and Goddesses* (New York: Routledge, 2002), 3.

28 Molefi Kete Asante and Abu S. Abarry, eds., *African Intellec-*

tual Heritage: A Book of Sources (Philadelphia: Temple University Press, 1996), 35–37.

29 On April 1, 1838, Darwin wrote to his sister Susan about his trip to the zoo. For more on Jenny the orangutan, see Jonathan Weiner, "Darwin at the Zoo," *Scientific American,* November 5, 2006.

30 Charles Darwin, *The Autobiography of Charles Darwin, 1809–1882,* ed. Nora Barlow (New York: W. W. Norton & Company, 1958), 144.

CHAPTER II: GENTLEMEN AND OLD PALS

1 Ibid., 2.

2 Janet Browne, *Charles Darwin: The Power of Place* (Princeton, NJ: Princeton University Press, 2002), 40.

3 Thomas Bell, quoted in Brian Gardiner's editorial in *The Linnean* 13, no. 4, 1997.

4 Charles R. Darwin and Alfred R. Wallace, "Proceedings of the Meeting of the Linnean Society held on July 1st, 1858," *Journal of the Proceedings of the Linnean Society: Zoology* 3. The printed publication is available at Wallace Online (wallace-online.org). The entire paper used Wallace's title (without crediting him specifically), but because Darwin was the first author, his name appears first on the title page and in all running heads.

5 The letter is missing, but Wallace summarizes his thoughts in his autobiography, *My Life: A Record of Events and Opinions* (London: Chapman & Hall, 1905).

6 Charles Darwin, *The Origin of Species* (London: John Murray, 1859), 488.

7   Richard Owen, "Darwin on the Origin of Species," *Edinburgh Review* (April 1860).

8   Thomas Henry Huxley, "Darwin's *Origin of Species,*" the *Times* (London), December 26, 1859.

9   For more on the X Club, see Ruth Barton, "'An Influential Set of Chaps': The X-Club and Royal Society Politics, 1864–85," *British Journal for the History of Science* 23, no. 1 (1990); and Browne, *Charles Darwin,* 2002.

10   See Leon Wieseltier, "A Darwinist Mob Goes After a Serious Philosopher," *New Republic* (March 8, 2013), and Thomas Nagel, *Mind and Cosmos: Why the Materialist Neo-Darwinian Conception of Nature Is Almost Certainly False* (New York: Oxford University Press, 2012).

11   For Huxley's complete 1889 essay "Agnosticism," see Thomas Henry Huxley, *Collected Essays,* vol. 5, *Science and Christian Tradition* (New York: D. Appleton and Company, 1902).

12   For more information, see Walter Kaufmann, *Nietzsche: Philosopher, Psychologist, Antichrist* (Princeton, NJ: Princeton University Press, 2013).

13   See Browne, *Charles Darwin* (2002), 104.

14   Max Müller, *Lectures on the Science of Language* (New York: Charles Scribner, 1862), 354. The lectures were delivered in April, May, and June of 1861.

15   Max Müller, "On the Results of the Science of Language," in *Essays Chiefly on the Science of Language,* vol. 4 of *Chips from a German Workshop* (New York: Charles Scribner's Sons, 1881), 449. This was originally delivered as a lecture at the Imperial University of Strasbourg, May 23, 1872.

16   Quoted in John van Wyhe and Peter C. Kjaergaard, "Going

the Whole Orang: Darwin, Wallace and the Natural History of Orangutans," *Studies in History and Philosophy of Biological and Biomedical Sciences* 51 (June 2015), 53–63.

17   Alfred Russel Wallace, "The Limits of Natural Selection as Applied to Man," *Contributions to the Theory of Natural Selection,* 2nd ed. (New York: Macmillan and Co., 1871), 370.

18   Ibid., 334–36.

19   Ibid., 344.

20   Ibid., 344–49.

21   Ibid., 334.

22   Ibid., 352.

23   Ibid., 359–60.

24   Shermer, *In Darwin's Shadow,* 161.

25   Darwin to Alfred Russel Wallace, March 27, 1869. Available from the Darwin Correspondence Project database at https://www.darwinproject.ac.uk/entry-6684.

26   For more on how spiritism emerged in Darwin's circle, see James Lander, *Lincoln and Darwin: Shared Visions of Race, Science, and Religion* (Carbondale: Southern Illinois University Press, 2010), 243–44.

CHAPTER III: THE DARK AGES

1   Müller, *Lectures on the Science of Language.*

2   Wallace, "The Limits of Natural Selection," 335, 359.

3   Charles Darwin, *The Descent of Man, and Selection in Relation to Sex* (London: John Murray, 1871).

4   Rudyard Kipling, "How the Leopard Got His Spots," in *Just So Stories* (New York: Doubleday, Page & Company, 1912).

5   Stephen Jay Gould, "Sociobiology: The Art of Storytelling," *New Scientist,* November 16, 1978.

6   This is how W. Tecumseh Fitch describes Darwin's theory in "Musical Protolanguage: Darwin's Theory of Language Evolution Revisited," *Language Log* (of the Linguistic Data Consortium of the University of Pennsylvania), February 12, 2009, available at http://languagelog.ldc.upenn.edu/nll/?p=1136. Darwin makes the comparison himself in *The Descent of Man,* 55.

7   See Darwin, *The Descent of Man,* 54.

8   Ibid., 83. This was an addition to the second edition.

9   Ibid.

10 Ibid., 68.

11 Ibid.

12 Ibid., 10. The earwigs were also an addition to the second edition.

13 Ibid., 77–78.

14 Ibid.

15 Max Müller, "Darwinism Tested by the Science of Language," *Nature* 1 (January 6, 1870), 256–59.

16 "Retrospect of Literature, Art, and Science in 1871: Science," *The Annual Register* (1871), 368. The name of the reviewer was never revealed.

17   For examples, see note 26. See also "Review of *Descent of Man*," *Athenaeum* 3 (April 1871), and "Review of *The Descent of Man*," *Edinburgh Review* (July–October 1871).

18   For more information about the Philological Society, see Fiona Marshall, "History of the Philological Society: The Early Years," available from www.philsoc.org.uk/history.asp.

19   Société de Linguistique de Paris. "Statuts de 1866, Art. 2." Available at: http://www.slp-paris.com/spip.php?article5.

20   See Barton, "'An Influential Set of Chaps.'"

21   Quoted in Paul C. Mangelsdorf, foreword to *Experiments in Plant Hybridisation* by Gregor Mendel (Cambridge, MA: Harvard University Press, 1965). This note was kept by one of Mendel's fellow monks, Franz Barina.

22   Theodosius Dobzhansky, "Nothing in Biology Makes Sense Except in the Light of Evolution," *The American Biology Teacher* 35, no. 3 (March 1973), 125–29.

23   See Morris Swadesh, "Sociologic Notes on Obsolescent Languages," *International Journal of American Linguistics* 14, no. 4 (October 1948), 226–35, and Stanley Newman, "Morris Swadesh (1909–1967)," *Language* 43, no. 4 (December 1967), 948–57.

24   Roger Hilsman, *American Guerrilla: My War Behind Japanese Lines* (Washington, DC: Potomac Books, 2005), 143.

25   Morris Swadesh, "Towards Greater Accuracy in Lexicostatistical Dating," *International Journal of American Linguistics* 21, no. 2 (April 1955), 121–37.

26   Edwin G. Pulleyblank, "The Meaning of Duality of Patterning and Its Importance in Language Evolution," in *Studies in Language Origins* (Philadelphia: John Benjamins, 1989), 1:53–65.

27 See "Celebrating the History of Building 20" on the MIT Libraries Archives, available at http://libraries.mit.edu/archives/mithistory/building20/index.html.

28 See Florence Harris, with James Harris, "The Development of the Linguistics Program at the Massachusetts Institute of Technology" (1974), on the website *50 Years of Linguistics at MIT: A Scientific Reunion, December 9–11, 2011,* available at http://ling50.mit.edu/harris-development.

CHAPTER IV: NOAM CHARISMA

1 Noam Chomsky, *Syntactic Structures* (The Hague: Mouton & Co., 1957), v.

2 Ibid.

3 Daniel Yergin, "The Chomskyan Revolution," *Noam Chomsky: Critical Assessments,* ed. Carlos Peregrín Otero (London: Routledge, 1994), 42.

4 John R. Searle, "A Special Supplement: Chomsky's Revolution in Linguistics," *New York Review of Books* (June 29, 1972).

5 Noam Chomsky, "The Case Against B. F. Skinner," *New York Review of Books* (December 30, 1971).

6 B. F. Skinner, "A Critique of Psychoanalytic Concepts and Theories," *Cumulative Record,* 3rd ed. (New York: Appleton-Century-Crofts, 1972), 238–48. In the same article he claims that Freud lets psychoanalysis "steal the show" from behavioral and environmental factors.

7 B. F. Skinner, *Verbal Behavior* (Cambridge, MA: B. F. Skinner Foundation, 2014), chapter 1 (e-book edition).

8   Noam Chomsky, "A Review of B. F. Skinner's *Verbal Behavior*," in Leon A. Jakobovits and Murray S. Miron (eds.), *Readings in the Psychology of Language* (Englewood Cliffs, NJ: Prentice-Hall, 1967), 142–143, available at https://chomsky.info/1967/.

9   Ibid.

10   Ibid.

11   Ibid.

12   Ibid.

13   The interview was originally published in *Omni* magazine's November 1983 issue. An online transcript is available at https://chomsky.info/interviews/.

14   Ibid.

15   Chomsly, "The Case Against B. F. Skinner."

16   Per Chomsky's March 10, 1984, letter to Lou Rollins, available at http://www.countercontempt.com/wp-content/uploads/2011/02/IMG.pdf.

17   Noam Chomsky, *Understanding Power: The Indispensable Chomsky,* ed. Peter R. Mitchell and John Schoeffel (New York: The New Press, 2002), 231.

18   Ibid, 245.

19   Noam Chomsky, "Comments on Dershowitz" (August 17, 2006), available at www.chomsky.info/letters/20060817.htm.

20   Noam Chomsky, "Reply to Hitchens's Rejoinder," *The Nation,* October 15, 2001.

21   Noam Chomsky, "Reply to Werner Cohn," *Outlook,* June 1, 1989.

22   Noam Chomsky, interview with Vince Emanuele for Veterans Unplugged, "Virtual Town Hall," December 2012. An archive of this interview is available at chomsky.globl.org.

23   Quoted in Tom Bartlett, "Angry Words," *The Chronicle of Higher Education,* March 20, 2012. The original quotation was from an interview in Portuguese: Da Redação, "Ele virou um charlatão," *Folha de S.Paulo,* September 1, 2009. (The title refers to Everett and translates as "He became a charlatan.")

24   Noam Chomsky, "A Special Supplement: The Responsibility of Intellectuals," *New York Review of Books,* February 23, 1967. www.nybooks.com/articles/archives/1967/feb/23/a-special-supplement-the-responsibility-of-intelle/.

25   Chomsky writes about this (and the others in "the prison dormitory") in "On Resistance," *New York Review of Books,* December 7, 1967.

26   See Harriet Feinberg, *Elsie Chomsky: A Life in Jewish Education* (Waltham, MA: Hadassah-Brandeis Institute, 1999).

27   For more on this history, see William I. Brustein, *Roots of Hate: Anti-Semitism in Europe Before the Holocaust* (New York: Cambridge University Press, 2003), and the United States Holocaust Memorial Museum's website at www.ushmm.org.

28   Jose Pierats, "The Revolution on the Land," in *The Anarchist Collectives: Workers' Self-Management in the Spanish Revolution, 1936–1939,* ed. Sam Dolgoff (New York: Free Life Editions, 1974). Pierats cites Augustin Souchy Bauer as the primary source for these numbers. See also Antony Beevor, *The Battle for Spain: The Spanish Civil War 1936–1939* (New York: Penguin Books, 2006).

29   From an interview in Noam Chomsky, *The Chomsky Reader,*

ed. James Peck (New York: Pantheon Books, 1987), 5. "Noam Chomsky," Britannica.com.

30   Paul Robinson, "The Chomsky Problem," *New York Times Book Review,* February 25, 1979 (a review of Chomsky's *Language and Responsibility*).

31   Eugene Garfield, "The 250 Most-Cited Authors in the *Arts & Humanities Citation Index,* 1976–1983," *Current Contents* 48 (December 1, 1986), 3–10.

32   Robin Blackburn, "For and Against Chomsky," *Prospect,* November 2005.

33   Jason Cowley, "Heroes of Our Time—The Top 50," *New Statesman,* May 22, 2006.

34   Larissa MacFarquhar, "The Devil's Accountant," *New Yorker,* March 31, 2003.

35   "Noam Chomsky," in Brian Duignan, ed., *The 100 Most Influential Philosophers of All Time* (New York: Britannica Educational Publishing, 2010), 314–16.

36   Counts come from Chomsky's website, www.chomsky.info/books.htm.

37   See Marc D. Hauser, Noam Chomsky, and W. Tecumseh Fitch, "The Faculty of Language: What Is It, Who Has It, and How Did It Evolve?" *Science* 298, no. 5598 (November 22, 2002), 1569–79.

CHAPTER V: WHAT THE FLYCATCHER CAUGHT

1   Daniel L. Everett, "Pirahã," in *Handbook of Amazonian Languages,* ed. Desmond C. Derbyshire and Geoffrey K. Pullum

(Berlin: Mouton DeGruyter, 1986), 1:200–326. For an example of Everett's academic adulation of Chomsky, see pages 256–57.

2　Daniel Everett, "Cultural Constraints on Grammar and Cognition in Pirahã: Another Look at the Design Features of Human Language," *Current Anthropology* 46, no. 4 (August–October 2005).

3　For more on Everett's personal history, see his interview in the *Telegraph* ("Daniel Everett: Lost in Translation" by William Leith, April 10, 2012) and a profile in the *New Yorker* ("The Interpreter" by John Colapinto, April 16, 2007).

4　Learn more about the Pirahã language and Everett's initial experience with the tribe in Daniel Everett, *Don't Sleep, There Are Snakes* (New York: Pantheon Books, 2008). In the book, Everett also recounts the experiences of earlier missionaries who were unsuccessful.

5　Everett addresses this in *Handbook of Amazonian Languages*.

6　Everett, *Don't Sleep, There Are Snakes,* 132.

7　Ibid.

8　Ibid.

9　Jennifer M. D. Yoon, Nathan Witthoft, Jonathan Winawer, Michael C. Frank, Daniel L. Everett, and Edward Gibson, "Cultural Differences in Perceptual Reorganization in US and Pirahã Adults," *PLoS ONE* 9, no. 11 (November 20, 2014).

10　Ibid.

11　Ibid.

12　Ibid.

13   Rafaela von Bredow, "Brazil's Pirahã Tribe: Living Without Numbers or Time," *Der Spiegel,* May 3, 2006.

14   Elizabeth Davies, "Unlocking the Secret Sounds of Language: Life Without Time or Numbers," *Independent,* May 6, 2006.

15   Liz Else and Lucy Middleton, "Interview: Out on a Limb over Language," *New Scientist,* January 16, 2008.

16   Quoted in Geoffrey K. Pullum, "Fear and Loathing on Massachusetts Avenue," *Language Log* (of the Linguistic Data Consortium of the University of Pennsylvania), November 29, 2006, available at http://itre.cis.upenn.edu/~myl/languagelog/archives/003837.html. Archives of the complete message, which was dated November 28, 2006, are available online, including on the Boston Area Neuroscience Talks group on Yahoo (http://yhoo.it/1SdpILf).

17   Bartlett, "Angry Words."

18   The 2007 article is still available on LingBuzz at http://ling.auf.net/lingbuzz/000411.

19   "Recursion and Human Thought: Why the Pirahã Don't Have Numbers," *Edge,* June 11, 2007, available at https://edge.org/conversation/daniel_l_everett-recursion-and-human-thought. In the "Reality Club" follow-up discussion, Pesetsky takes issue with Everett's claims that Pesetsky and his coauthors have ties to MIT: "We are all experienced researchers, and we are not all from MIT."

20   MIT keeps track of all dissertations and advisers in a public online database called DSpace@MIT (dspace.mit.edu).

21   Ibid.

22   Colapinto, "The Interpreter."

23   Andrew Nevins, David Pesetsky, and Cilene Rodrigues, "Pirahã Exceptionality: A Reassessment," *Language* 85, no. 2 (June 2009), 355–404.

CHAPTER VI: THE FIREWALL

1   Everett, *Don't Sleep, There Are Snakes*.

2   "Excerpt: 'Don't Sleep, There Are Snakes,'" from the series *Best Books 2009,* December 23, 2009, available at http://www .npr.org/templates/story/story.php?storyId=121515579.

3   Everett, *Don't Sleep, There Are Snakes*.

4   Quoted in Bartlett, "Angry Words."

5   Quoted in Claudio Angelo, "O Iconoclasta" ("The Iconoclast"), *Folha de S.Paulo,* February 1, 2009.

6   Michael Tomasello, "Universal Grammar Is Dead," *Behavioral and Brain Sciences* 32, no. 5 (October 2009), 470–71. From the article's abstract: "To make progress in understanding human linguistic competence, cognitive scientists must abandon the idea of an innate universal grammar and instead try to build theories that explain both linguistic universals and diversity and how they emerge."

7   Vyvyan Evans, *The Language Myth* (Cambridge, UK: Cambridge University Press, 2014), i.

8   Larry Trask, quoted in Andrew Brown, "A Way with Words," the *Guardian,* June 25, 2003. Trask died in 2004, a year before Everett's paper "Cultural Constraints on Grammar and Cognition in Pirahã" was published.

9   Author's phone interview with Noam Chomsky on May 3, 2016.

10   Noam Chomsky, "What Is Language and Why Does It Matter?" Lecture given at 2013 Linguistic Society of America Summer Institute at the University of Michigan, available at https://www.youtube.com/watch?v=-72JNZZBoVw.

11   Rachel Feltman, "Birdsong and Human Speech Turn Out to Be Controlled by the Same Genes," *Washington Post,* December 11, 2014.

12   Marc Hauser, et al., "The Mystery of Language Evolution," *Frontiers in Psychology,* May 7, 2014.

13   Chris Sinha, "Language and Other Artifacts: Socio-Cultural Dynamics of Niche Construction," *Frontiers in Psychology,* October 20, 2015.

14   Andy Clark, *Being There: Putting Brain, Body, and World Together Again* (Cambridge, MA: MIT Press, 1997), 193.

penguin.co.uk/vintage